beautiful minds

BEAUTIFUL MINDS

THE PARALLEL LIVES OF
GREAT APES AND DOLPHINS

MADDALENA BEARZI
& CRAIG B. STANFORD

Harvard University Press

Cambridge, Massachusetts & London, England ≈ 2008

A Caravan book. For more information, visit
www.caravanbooks.org

Library of Congress Cataloging-in-Publication Data

Bearzi, Maddalena.
  Beautiful minds : the parallel lives of great apes
and dolphins / Maddalena Bearzi and Craig B.
Stanford.
      p. cm.
  Includes bibliographical references and index.
  ISBN 978-0-674-02781-7 (cloth : alk. paper)
1. Apes—Behavior.   2. Dolphins—Behavior.
3. Animal societies.   4. Psychology, Comparative.
I. Stanford, Craig B. (Craig Britton), 1956–   II. Title.
QL737.P96B39 2008
599.88'1513—dc22   2007046199

Maddalena: *For Charlie*

Craig: *For Erin*

CONTENTS

beautiful minds

BEAUTIFUL MINDS

ONE HAS HANDS much like yours or mine and can use them to skillfully manipulate a tool, delicately groom a partner, or speak in fluent sign language. The other doesn't have hands at all. One looks like you or me, more or less. The other has the body of a cruise missile. One swings through the trees of an African forest; the other dives deep in cold oceans. Great apes and dolphins would seem to have very little in common. They live in worlds so different that you would have to dissect one to find that their organs and limbs share any common features. They are both mammals, but distantly related; the two groups haven't had a common ancestor in nearly a hundred million years. A gorilla and a bottlenose dolphin are about as closely related as a mouse and an elephant.

In spite of these differences, dolphins and apes—and by extension ourselves—share some strikingly similar and profoundly important traits. All three groups—the various dolphin species, the four great apes, and we humans—possess the acme of brains on Earth today. With due respect to a few other brainy animals like elephants, the cetaceans—dolphins and whales—and higher primates are the most cerebral of the world's creatures. We are all highly intelligent relative to the millions of other co-inhabitants of Earth. We live in highly complex, often fluid societies which defy the easy categories that apply to most other animals. The two creatures have evolved in parallel, exemplifying what biologists call convergent evolution. Although a casual observer won't see these parallels, research on dolphins and apes has produced increasingly abundant evidence for the comparison.

This is why we have written a book about apes and dolphins. Although both have been the subject of many other books, rarely have the two been considered as companion species, evolutionary partners in ways that are not immediately apparent to most people. And we two—Stanford, a primatologist, and Bearzi, a dolphin biologist—felt a consideration of the surprising parallels between these two creatures might result in some timely lessons for humankind. We decided to write the book in the first person, so each of us could be the voice for the animals we have spent our lives observing.

However you define intelligence, apes and dolphins are second only to humans in brainpower. Their brains are enormous in comparison to the size of their bodies. This brainpower has allowed dolphins and apes to possess communication skills and social interactions so complex that we are only now beginning to understand how they work. Unlike most animals, apes and dolphins tend to live in flexible, open societies, and the relationships among individual animals are based on long-term memory of who is whose friend, and who owes whom a favor.

This combination of intelligence and social complexity is incredibly rare. It occurs on Earth today mainly in the living tips of the two great lineages: the cetaceans (dolphins and whales) and great apes (chimpanzees, bonobos, gorillas, and orangutans). In this book we suggest that the parallels between great apes and dolphins point us toward a deeper understanding of what it means to be human. We are, after all, the ultimate in big, brainy social complexity on the planet and possibly in the universe. We share with apes practically one-hundred percent of our evolutionary baggage, so our intellect is at its most basic primate level the same as theirs. Dolphins are an entirely different story; their big brains evolved without any historical connection to our own. But in this book we will consider whether the reasons for dolphin intelligence and social complexity have parallels in the great apes' and therefore our own evolution.

Apes

They sit like worried buddhas, their brows furrowed and eyes burning from underneath, just a few inches away. Some are handsome and virile, others are youngsters, and a few are old and shabby—just like the people gathered around me outside the zoo exhibit. A strikingly handsome male chimpanzee is grooming himself, looking regal compared to kids around him. An old female, Pandora, has a ragged body capped by a massive pink swelling that hangs off her posterior end. I find it pretty hideous, but its appeal apparently goes way beyond the human aesthetic range, because she raises excitement among the male chimpanzees wherever she goes.

The glass barrier between us looks at first glance like it is there to protect us from savage apes, but I know better. It's also there to protect them from savage, disease-ravaged visitors. Chimpanzees can catch virtually all our diseases, for the diseases evolved to attack us have only a tiny adjustment to make to invade a creature so genetically like us. With thousands pressing their noses to the glass every day to connect with the mind of an ape, chimps without the barrier would be like lost tribes at the mercy of the germs carried to them by well-meaning missionaries.

The zoo has gone to great lengths to educate the public about the plight of the chimpanzee; its endangered status in Africa, the rate at which it is being hunted out of existence to feed people even while its forest homes are being logged. It's all happening with breathtaking speed, several million years of evolution being scratched out in an instant. But the message seems lost on the zoo visitors this morning; they're just trying to make a connection with an ape.

One of the children hoping to leap momentarily into the skin of an ape is my son. Adam looks through the window that separates us from the chimpanzees and then glances back at me. At seven years old, he stands about the same height they do. He's fascinated by all primates, partly because people naturally are, and also because his father is always talking about them or flying off to some remote place to study them. He was recently mentioned in a newspaper story about my work with chimpanzees—the journalist had watched Adam watching the chimpanzees in the zoo just as he is today and found it an irresistible hook for the story.

Adam's understanding of our connection to chimpanzees is simple but accurate. We are they, and they are us, for the most part. Adam knows that people "came from" apes long, long ago. Like many kids, he knows all about dinosaurs and has no trouble imagining that there may have been myriad animal forms in Earth's history like nothing we have seen since. It's usually adults who avoid acknowledging the continuity between them and us. Adam sees that the chimps are not monkeys, even though that is what most zoo visitors around us apparently believe. Chimpanzees are, in fact, more closely related to humans than they are to gorillas, all hirsute, beetle-browed appearances to the contrary.

These chimpanzees, born and raised in the zoo, seem almost domesticated. Their physique is like that of bodybuilders compared to their wild counterparts. At four years old, zoo apes are the size that ten-year-old chimpanzees would be in the wild. With an abundant, carefully managed food supply, their growth rate is astounding, their muscles bulging at an early age. Yet zoo chimps are different from their wild-living cousins in many ways, including the ways in which they relate to one another.

Despite their smaller size, in the wild, chimpanzees are altogether different animals. They are sinewy and immensely powerful. I have seen males break off saplings to drag about in their macho charging displays at other males. Once a particularly bellicose male broke a tree on top of me in an apparent attempt to bully me. It was highly successful. Just as males can be brutal bullies, females are devoted, loving mothers. But females also exercise their own power, ganging up at times on unruly males. Despite their strength, I am always struck at how restrained most chimpanzee behavior is; they reserve their brute power for those rare times when it is truly needed. They spend nearly all of their lives peacefully plucking fruits, grooming one another, and sleeping; only in the smallest fraction of their lives do they let loose and show their power.

Chimpanzees are one of four species of great apes, along with bonobos, gorillas, and orangutans. Their anatomy is close enough to ours that they belong in the same taxonomic family. Only because the earliest classifier, the Swede Carolus Linnaeus, was devoutly religious did the apes end up in their own family, separate from our own. Several million years ago, they and we shared the same ancestor. Then, the evolutionary lines split. Our own direct ancestors eventually stood up and walked, their brains mushroomed, and sophisticated intelligence blossomed. So the apes offer us a window onto whom we once were, not so long ago.

Jane Goodall began her famed study of wild chimpanzees in 1960, and not until the mid-1970s did a clear understanding of chimpanzee society emerge. The complexity of chimpanzee social behavior is such that it took nearly two decades to begin to understand how their society works. And now in spite of nearly half a century of research, we still have major gaps in our understanding about these apes. They are smart and resourceful, sharing the same range of emotions as ourselves. In watching wild chimpanzees, I have seen fear, aggression, and nurturing, but also guilt, shame, and love. This is not just anthropomorphism on my part; given our kinship, our *a priori* assumption is that the emotional motivations of great apes are similar to our own.

The great apes live in the tropical forests of Africa and Asia. Their plight in the wild is so severe—from habitat loss to poaching for meat to rare viral epidemics—that they may well all disappear from Earth within our lifetimes. There are today an estimated 200,000–250,000 great apes of all four species combined—most of these are chimpanzees—and their numbers are in freefall. While they are still with us, they have much to teach, if only we would observe and listen.

How genetically similar are we to the great apes? By some estimates, the percentage of spots on our mutual DNA sequences that do not match up is substantially less than one percent. Now that the human genome map is more or less complete, molecular biologists have turned to mapping the chimpanzee genome. When that work is complete, scientists may begin to understand not just the percentage of similarity between us, but which genes differ in function. We are beginning to crack the code.

Dolphins

In the cold ocean waters of the world swim intelligent torpedoes. If people and apes are cousins, then people and dolphins are long-lost relatives. We are both mammals; we nurse our babies, we have hair (precious little in the dolphins' case), we have big brains and live in complex societies. But in most respects, notwithstanding the mammalian link, they are perhaps as different from us as any smart extraterrestrial creature would be. Their world is seawater, not land, and they navigate their territory by sonar rather than by sight. Buoyancy, not gravity, defines their body form and function. Where we expect legs, they have flippers. Their vocal communication, clicks and whistles, is unfathomably unfamiliar to us.

Recently my children and I stood watching dolphins cavort in a large aquarium tank. Their beauty comes from grace and power, artfully blended. But whereas a child will immediately see the human connection when staring at a chimpanzee, there is no such mental leap when watching a dolphin. "What are they related to?" my son asked. He knows that these are mammals and not fish. His question is rather, "Where did they come from?" This is not a question we would ask about an ape, because we already know the answer. We came from apes, and apes came from monkeys, literally. But dolphins? When I start talking about dolphins and whales evolving over thousands of generations from animals that walked about on land, and over millions of years migrated into the sea, I can see my son trying to wrap his mind around the idea. He doesn't dismiss it, as some theologically driven folks might, it's just an idea that is a bit too abstract for him right now. But looking at an ape, there's only a baby step of faith needed to see that its history is our history, until very recently. The most recent evidence from fossils and from DNA indicates a fork in our shared ancestral paths about 6 million years ago; all earlier forms of people, from Lucy to Neanderthals, have lived since that split.

But dolphins are very distant relatives. Many things about us that we consider human we see in great apes, but those same qualities seem to be lacking in dolphins. Our range of facial expressions, for example, which show so much of our collective soul, are reciprocated by the frozen faux-smile of the dolphin. Our multi-limbed, three-dimensional bodies, so apelike, contrast with the dolphin's sleek, sluglike frame. The way the two species express themselves could not be more different. Many of a dolphin's sounds, even those that our ears can detect, are produced not in a larynx but deep inside the skull.

We have each spent large parts of our lives watching the animals we write about; Stanford studying great apes and Bearzi studying dolphins. We are utterly fascinated by them, we feel passionately about them as fellow creatures, and we hope to convince you that understanding the unlikely parallels between these creatures gives us a rare glimpse into the origins of that most human of all qualities, our intellect.

chapter one

AN ETERNAL FASCINATION

IT IS SIX O'CLOCK. The aquarium's doors are now closed to the public, leaving me alone while I wait for a friend in an interview. An assistant whispers that I can look around for half an hour. Among the jellyfish, shark, and sea turtle exhibits, I am drawn toward one of the lateral portholes that look out into the submerged world of Miko and Mara, a pair of elderly dolphins kept here in captivity. The tank seems empty, silent.

Out of the blue, I see a gray body slowly rising from the bottom. Nose glued to the porthole, I strain to follow the fluid movements, looking downward until it almost hurts. Suddenly, an inquisitive eye pauses just inches from mine. It's only for an instant. Then, as though it were never there, the dolphin continues its rise to the surface and vanishes into the gloomy water. With my heart thumping, I frantically clean the winter sludge from the glass, hoping for another encounter.

And here she is once again. It's Mara. I can see her sex as she approaches my little window in slow motion. She lingers this time, inspecting my face, so different from hers, so alien. With one eye, then the other, she rolls, seemingly weightless, in a sort of circular dance. An intense shiver runs down my spine. Three, four, five times we meet and meet again, so close and so distant, divided by that glass. My friend calling my name brings me back to the real world. His interview is over and now he's ready to leave. I reluctantly pull myself away from the window and turn to go. As I walk away I feel

Mara's eyes following my every step until I have disappeared behind the aquarium gate.

I still don't know what it was that struck me most: her living sad glance looking straight into my eyes, her mysterious "smile," or that harmonious, lethargic ballet. I don't really remember if it was in that moment or some other that I decided to become a field biologist, but I know that I felt at home with that dolphin and suddenly realized how caged and confined her world must be. What I do remember is forming a firm conviction that dolphins have the right to their own home.

My brief encounter with Mara gets lost among many stories and legends that tell of dolphins; intelligent, beautiful, curious, sensitive, playful dolphins. Since the dawn of civilization, humans have been fascinated with these elusive marine mammals and their mysterious existence beneath the oceans. To us, there is something inexplicable about dolphins, about their life and their world.

Signs of the magical and strong bond between humans and dolphins can be found engraved in the prehistoric caves of the Pyrénées. Respect for these creatures was demanded in ancient Egypt, and the people of Greece regarded the killing of a dolphin as a sacrilege against the gods, a horrible, perfidious act punishable by death.

Intelligence, consciousness, and compassion were among the words used by the ancient Greeks to describe their "companions of the sea." The philosopher Plutarch regarded them as the only creatures that seek friendship for purely altruistic reasons. Greek myths portray that culture's fascination with dolphins: the sun god Apollo assumed the form of a dolphin when he founded his oracle at Delphi on the slopes of Mount Parnassus, and Orion was saved from drowning by a sociable dolphin and carried into the sky, riding on his back. Then there is Dionysus, the Greek god of wine and mirth, who once booked passage on a ship from the island of Ikaria bound for Naxos. The ship's crew was actually a band of pirates posing as merchant sailors, whose secret plan was to capture their passengers and sell them into slavery. When Dionysus discovered their conspiracy, he used his divine powers to punish them by causing the ship's mast to sprout branches, the men's oars to become snakes, and a strange flute to play. To escape this divine madness, the pirates cast themselves into the water, where the ocean god Poseidon changed them into dolphins and commanded them to serve humankind forever.

A century later, the Roman philosopher Pliny the Elder told the story of a peasant boy living on the shores of the Mediterranean Sea who befriended a solitary dolphin named Simo. Every day, Simo used to take the boy on his back across a stretch of water between the child's home and his school. Tragically, the boy fell sick and died. Upon the boy's death and for many days thereafter, Simo kept returning to the place of their meeting until he too died of a broken heart. Simo's story is one of the countless legends of friendship between children and dolphins that were popular during the Roman Empire. As it was with the Greeks, the Romans were fascinated with these creatures.

These are some of the many legends that kept my adolescent mind awake late into the night, eager to read more about these wonderful and somehow magical dolphins. One story after another, one legend overlapping the next, until my dreams became filled with gods, seas, and dolphins, and from time to time, I would find myself among them.

The mystical relationships with these marine mammals proliferated in the annals of ancient Greek and Roman culture, but there are other cultures throughout the world that have honored dolphins and whales, many of which still exist today. Among the most marvelous folklore is that of Australian aborigines who tell of being in communication with bottlenose dolphins in the Indian Ocean for thousands of years. They have a medicine man that calls and talks to them telepathically, and through these communications guarantees the good fortune and happiness of the tribe.

No less enthralling are the tales from the banks of the Amazon River in Brazil, where some tribes believe that the boto, a local species of freshwater dolphins, hold the power to transform themselves into beautiful young men in order to seduce women during celebrations and ceremonies. Their belief is so strong that some children in the tribe are thought to have been fathered by these animals.

It's hard to draw a line between legend and reality when talking about the human fascination with dolphins. Stories abound of animals rescuing shipwrecked sailors, saving struggling swimmers by fending off shark attacks, of children riding on the backs of wild dolphins. Dolphins have become our saviors in art, poetry, and literature—even on television. Who can forget Flipper (or rather one of the five actual Flippers), the benevolent and watchful "puppy of the sea," soliciting human companionship and protecting all that is good? It seems even dolphins can attain Hollywood stardom.

Some of the legends told by our ancestors were not so far from reality. Consider for a moment interactions between solitary dolphins and humans. Documented cases of these relationships appear more and more frequently in recent times. In the Italian port of Manfredonia lives Filippo, a lone and sociable adult male bottlenose dolphin. Filippo spends his time floating next to his favorite moored boat or playing with the propellers of other vessels coming into and leaving the port. Filippo is anything but active. Unlike a wild dolphin, whose days are spent foraging sometimes over vast distances, Filippo spends his time leisurely catching easy fish inside the port, masturbating with inflatable dinghies, but more than anything else, he loves interaction with humans, above all women, whom he approaches demanding some kind of sexual contact. In the company of men, Filippo prefers to play macho games, sometimes becoming a bit aggressive and intolerant, striking out with his beak in anxious moments. Filippo is one of the greatest examples of a human–dolphin relationship. It is easy to find many similarities between his dolphin and our human behavior, whether we care to admit it or not.

A few years after my first encounter with Mara at the aquarium in Italy, I finally found myself at home with dolphins. This time, it was in their home. Working at sea with dolphins had become my job as a scientist, my greatest passion, and a natural part of my everyday life. Spending hours, days, and years observing their behavior, following their movements, recording their sounds, I learned that they were not the big-brained Einsteins, wizards, or philosophers as described in so many fables, but rather they were very intriguing, highly complex and flexible social animals completely adapted to an ocean life—one that is still little understood and rich with their many secrets. The more I came to know them, the more I became skeptical of the chronicles and folklore depicting dolphins as creatures who voluntarily associate with people. The more I learned about their existence, the more I realized that the relationship between humans and sociable dolphins has not always been as congenial as I thought. Today more then ever, we are walking a thin line in our connection with these animals.

It is true that dolphins are social by nature. But it is also true that dolphins do not always welcome physical contact with humans, and their existence as well as their home require as much respect as we might give to our own. My scientific vision and my affection for these animals were pulling me away from that need for physical contact that we humans impose on creatures that seem to fulfill our criteria of personhood. And with these realizations, I closed the door on the stories of humans and dolphins that had filled my adolescence . . . until one chilly winter morning.

I was following a school of bottlenose dolphins in the coastal waters of Los Angeles, as I do on a weekly basis. This particular morning, my "metropolitan" dolphins were foraging for prey somewhat sluggishly, moving along the beach several hundred meters from the surf line. I was working to photo-identify the school, trying to take clear pictures of all the dorsal fins of all the individuals in the group. This is a technique used by many cetologists (scientists studying whales and dolphins) to recognize specific individuals. Not unlike a human fingerprint, a dolphin's dorsal fin works well for identification thanks to its distinctive scars and notches.

Since I began my study of these animals, they had never seemed to mind my presence, often riding the bow waves created by my boat moving through the water in a behavior called bow riding or just continuing their natural behavior with me at their side as a tolerated intruder. I would like to believe that in some peculiar way, my boat and its researchers, jam-packed with cameras, hydrophones, and computers, had become part of their everyday life's scenery.

The morning was foggy and cold, and the dolphins were moving north and then south, stopping here and there for short inspecting dives. I was following and recording their activities as they encircled a large school of sardines just off the Malibu pier. The fish were trapped ingeniously, as if in a net shaped by a tight formation of nine dolphins. Just after they began feeding, one of the dolphins in the group suddenly left the circle, swimming offshore at a high speed. In less than an instant, the other dolphins left their prey to follow their companion. This was an odd behavior for my metropolitan dolphins. Usually they moved back and forth very close to the beach, taking the time to entirely deplete the school of fish on which they were dining while occasionally milling at the surface like a bunch of oversized floating buoys. To abruptly stop feeding and take off in an unrelated direction was rather peculiar.

Always curious, I left the schooling fish, still visible from the surface, and accelerated into the incoming waves to follow the dolphin group. We were at least three miles offshore when the dolphins stopped suddenly, forming a large ring without exhibiting any specific behavior. That's when one of my assistants spotted an inert human body with long, blonde hair floating in the center of the dolphin ring. Breaking the ring, I maneuvered the boat closer to the girl and asked if she was okay, but she looked at us with no apparent response. We decided to get her into the boat. As we came closer, she raised a weak arm in what seemed a plea for help. Her face was pale and her lips were blue as I pulled her fully dressed and motionless body from the water. We called the local lifeguards on the radio and were told not to touch anything until they arrived, which we promptly disregarded, helping her out of her wet clothes and trying to get her warm by using blankets and contact from our bodies. When we pulled her from the water, she was hypothermic. She began to respond and as we turned to go back to port I noticed that the dolphins were gone.

Later at the hospital, a doctor told me that she was from Germany, on vacation in Los Angeles. She was eighteen and was evidently trying to swim offshore to die by suicide. When we found her, she had all her travel and identification documents tied around her neck in a plastic bag. The bag also contained a letter, which I suspect must have been some explanation for her decision to kill herself. The doctor told me that had we not found her when we did, she would surely have died. That is when I remembered the dolphins and how we had come upon her.

Sheer coincidence? Perhaps. But I still think and dream about that cold day and that tiny, pale girl lost in the ocean and found again for some inexplicable reason, by us, by the dolphins? My adolescent drawer filled with anecdotes and legends is still closed, but after that chilly morning, my fascination remains acute.

What is it that makes dolphins so appealing to us? I often dwell on that question. Sometimes it is hard to be strictly scientific and deny the Greeks' model of an intelligent, conscious, and compassionate creature. And so it seems tough to ignore the many similarities that dolphins share with humans. They are, it would seem, more like us in their family bonds, the care and education of their young, culture, politics, social structure, even the ability to be sensitive and emotional. And they rival our capacity to experience fear, pleasure, and pain.

How can I forget the heartbreaking expression of Mara, enslaved in those four walls of the tank? Was it just my vivid imagination that her lethargic behavior and her soulful expression reflected her memory of life in the open ocean, a life without barriers and limits? Was she choking on the stale and confined waters of her tank? Did she miss her companions? Or was I just trying to humanize that dolphin and her feelings?

What about Filippo and his strong and sometimes too passionate feelings toward our species? Also the fruit of a scientist's imagination? Or is his behavior an attempt to shift his emotions from his missing companions toward these terrestrial, biped friends?

Communication may constitute yet another important facet of our fascination with dolphins. I can't count the number of times people have asked me, "Can we talk to them?" The origins of this mindset probably came from an extravagant scientist named John Lilly and the work he did with dolphins in captivity. Lilly believed that dolphins possess a highly developed brain and sophisticated linguistic abilities. His work captivated the public interest in the 1960s and his books became best sellers. He believed that humans and dolphins could communicate with each other, and his work was focused on breaking through an interspecies language barrier. Whether Lilly's work was convincing or whether it was our human desire to believe him, the idea that dolphins were intelligent mammals capable of communication became widely known. Whatever Lilly or we actually believe about human and dolphin communication, it remains that the story of the talking dolphin, especially for young children, today represents one of the most wonderful tales on Earth.

Next to communication, our fascination with dolphins may begin with our childlike attraction to their playfulness. They seem to experience great joy in playing with their companions or with animate or inanimate objects that they come across. They are often observed amusing themselves with balls in captivity as well as with plastic bags or strings of kelp in the open sea. They may toss a fish in the air with their beak and then grab it again and again as if in a silly mood, or engage in spectacular acrobatic jumps, lateral breaches, synchronous bows, and outrageous leaps. They may tease a sea turtle by pulling on its tail or imitate the movement of a small shark by swimming in a side-to-side motion. Joy seems to be contagious among dolphins.

Our fascination with dolphins may also derive from the particular shape of their mouths, forming what seems to us to be an indelible and friendly smile, even if, from time to time, that smile may conceal an upset or angry animal. Dolphins may become irritated and unpredictable, just as we do. Our friendly Filippo, for instance, cares for human contact more than anything, but now and again, he falls into a grumpy mood, displaying annoyance and sometimes even biting. Whatever we may think of his smiling face, in those moments, he isn't happy or friendly.

Are we drawn perhaps to the power of their streamlined bodies, able to streak through the water at high speed just as racing cars might on the open road? Or might it just be that the curiosity of a dolphin in many ways is so similar to the innocent inquisitiveness of a child?

Maybe it is because their world seems so different from our own; the multi-dimensional ocean world in which a dolphin seems to live in perfect harmony. If you have ever seen a dolphin leaping from the water or surfing the waves at the shoreline, it may have left you astounded at its almost effortless ability to move as it pleases in an aquatic medium so foreign to us. In the sea, our efforts seem feeble by comparison.

Perhaps our fascination with dolphins isn't really linked to any specific attribute. Maybe it is just the magnitude of what we still don't understand.

Almost Like Us

The gorilla in front of me sits like a Rodin statue, his chin propped in his hand. He is lying calmly in the deep shade of the forest, his broad, black back gleaming. Ruchina is a young adult male, a blackbuck, so-called because the saddle of silver hair has just begun to spread across his torso. He sits only a few yards from me, having decided to take his mid-day rest in this spot. He made a grand entrance a few minutes earlier, charging through a thicket, grabbing at a sapling tree, and cracking it in half before coming to rest like a runaway sixteen-wheeler. The rest of his group is fifty yards away, happily munching a salad of tender plants. The morning in this African forest is already warm, and sitting in the shade is the best place for both gorillas and their researchers.

Ruchina is something of an outcast, albeit self-imposed. As he approached adolescence, he began to find himself at odds with Zeus, the group's silver-back. Over time, he distanced himself from the rest of the group until he found himself in his present state, tagging along behind them like a child eager to be independent but ultimately tied by a psychological leash to a parent. Meanwhile, he is frustrated at the lack of attention he receives from females and his lack of status in life gener-ally. So we have to take care, because Ruchina is prone to vent his anger on the people who follow him around. This day, he got rid of his negative en-ergy merely by killing a small tree.

As he sits near us, occasionally turning an eye in our direction, he seems the epitome of calm. But it can change at a moment's notice if someone moves too fast or steps in the direction he was planning on moving. So we watch and admire carefully. Everyone who has sat next to an ape and stared into his eyes must ponder the same question. What do they understand? What goes on inside their heads? We swing between anthropomorphic thoughts of their essential humanness and denial of their intellect altogether. This is because of the intellectual black box they represent. As we shall see, fascinating recent research is opening the box for us a bit, and if interpreted correctly, will shine a whole new light on intelligence and cognition in gorillas and the other great apes.

Our ignorance of apes, and of their common bonds with ourselves, has spurred our fascination with them and guided much of the history of inquiry into their lives. While most of the dolphin's life is hidden beneath the waves, the apes conceal themselves in tropical forests that until recently were too remote for the adventures of even the most intrepid explorers and scientists. And we have spent much of history trying to deny that they and we are the closest of kin. Richard Owen, a renowned British paleontologist of the mid-nineteenth century, wanted to reclassify humans in an entirely different subclass of mammals than the apes. This would be like deciding that horses and zebras should not be considered horses simply because zebras have stripes. Owen was driven by a religious conviction that people and apes could not share a recent, common ancestry. But after seeing an orangutan in the London Zoo, Owen's contemporary and famed naturalist Charles Darwin wrote that people probably "retained from an extremely remote period some degree of instinctive love and sympathy."

Even a century later, the degree of kinship between people and apes was still being challenged and widely misunderstood. Many scientists studying the fossil record had long believed that apes and humans had not had a common ancestor in 15 million years or more. Then in the late 1960s, the biochemist Allan Wilson and Vincent Sarich, then an anthropology graduate student at the University of California, Berkeley, compared the immune system similarities of humans and apes. They devised a scale of the strength of antibody reactions in relation to the likely evolutionary distance between the species. Their conclusion: that people and chimps last shared a common ancestor only 5 million years ago. The idea of such a recent common ancestry outraged many in the scientific community, not to mention the public. But as further genetic research showed, Wilson and Sarich were just about right. Our kin and we were one and the same only several million years ago.

Chimpanzees are familiar to all of us; they serve as the basis for all of our imaginings about what human ancestors may have been like millions of years ago. They are, after all, our cousins. Chimpanzees live nearly human life spans, up to sixty years in captivity and forty-five or more in the wild. Their life cycle is almost human; a several-year period of infant dependency, reaching puberty in the early teen years. Females begin to have offspring at around fifteen and, after an eight-month pregnancy, give birth every four to five years for the rest of their lives.

We recognize them as our kin because they are so much like us physically, beyond the hair and the four-legged posture. And the physical similarities are not as riveting as the psychological, social, and emotional similarities. Apes have the same psychosocial needs as children, from everything we understand about them. They can learn language skills, but only when reared in a properly social environment with nurturing love and attention, since their development is hinged on their emotional needs, just like children.

Through the centuries humanity has regarded apes as heroes, enemies, or victims depending on the historical attitudes of the moment. In ancient times apes were considered low, derogatory creatures and a symbol of malice. They were ruthlessly trained for human amusement with little or no respect and were rumored to possess mystical powers that could be utilized in potions or spells: in some traditions the urine of an ape near the door of an enemy would make that person hated, and an ape's eye had the power to render its possessor completely invisible.

In ancient Egypt the reputation of apes changed for the better. Baboons were thought to represent the sun, moon, and stars and were regarded with great respect. Apes were highly cherished and loved. The ancient Egyptian practice of mummifying the dead often included apes, which were considered sacred pets and prepared as mummies to serve their masters in the eternal afterlife. By the time of the Old Kingdom, which began around 2650 B.C., baboons were closely associated with Thoth, the god of the moon, wisdom, and sacred writing. Thoth was either portrayed in human form with the head of an ibis or as a baboon with a furled and pensive brow. The Nubian tribes who inhabited the area that is now Egypt and Sudan believed that apes could understand human speech and their ability to learn a language exceeded that of children.

Ancient India was another good place to be a primate. In the Ramayana of Valmiki, the first epic poem written in Sanskrit recounting the mythical tales of King Rama and his travels, the powerful Vanaras were a monkey-like species of warriors loyal to King Rama. They were rumored to be able to converse with each other and served as the right hand of Rama in his battle against the demons. The most powerful of the Vanaras was Hanuman, half monkey, half god, who became the symbol of loyalty, strength, and self-sacrifice. He possessed godly powers enabling him to fly and change his physical aspect. Hanuman could become gigantic, tiny, invisible, or similar to a man. He was so strong as to shake the mountains, demolishing the cliff tops as he flew.

Chinese and Tibetan legend speak of the great stone ape Sun Wukong, who was born from a rock with supernatural powers and rose quickly to become the king of the monkeys on Earth. Mischievous and egocentric, Sun Wukong lived peacefully until realizing to his dismay that he was mortal like other monkeys. In a quest for immortality, the Monkey King wandered the world in search of someone able to teach him the secrets of perpetual life. He eventually enlisted the help of a great Taoist sage who, among other things, uttered the words of illumination and taught him the secret of seventy-two transformations and how to fly among the clouds. With a powerful magical weapon that he stole from the Dragon King, Sun Wukong battled many demons in service of the ruler of the universe, the Jade Emperor, until his ego again got the better of him and he left heaven in disgust. The infuriated Jade Emperor dispatched a powerful army to obliterate Sun Wukong, but the stone ape proved too powerful and rebuffed the attack. The Emperor then recruited the help of the Buddha, who was able to subdue Sun Wukong with a simple wager. After centuries of service to the heavenly empire, Sun Wukong finally made peace with the Jade Emperor and his court, gaining the title of "Buddha Victorious against Disaster." This great stone ape was even considered (though not eventually chosen) as a possible mascot of the 2008 Olympics in Beijing.

It was Christian symbology that once again changed the ape's destiny from that of venerated god to a symbol of vanity and greed. Some legends tell of Satan recreating himself in the form of an ape. Western culture began to regard apes as immoral and a representation of insecurity and immodesty, the lowest form of our bestial nature.

It is not clear from these legends and fables whether they relate to apes or monkeys or both.

The distinction between apes and monkeys was not made or understood in ancient times. It is even difficult to precisely trace the taxonomy of these species because the terminology has changed several times throughout the history of the study of primates.

The mention of apes in print dates back to occasional references in the twelfth century, but it is not until the beginning of the twentieth century that they emerged as subjects of popular novels like *Tarzan of the Apes* by Edgar Rice Burroughs, the legendary 1914 story of a boy lost in the jungle and raised by apes. Four years after its successful publication, the silent movie version of the story starring Elmo Lincoln appeared in theaters to standing ovations, opening the door to the ape movies of the future. The 1933 film *King Kong* began with the words of an Arabian proverb: "And lo, the beast looked upon the face of beauty. And it stayed its hand from killing. And from that day, it was as one dead." In what became one of the most popular movies ever made, the fifty-foot-tall gorilla King Kong introduced moviegoers to an ape personality that at one moment was savage and at the next tender and protective. It became time for monkey business in show business. After the immense success of *King Kong*, Hollywood went ape-crazy, producing an average of a monkey movie every two years or so. From the 1949 *Mighty Joe Young*, in which a ten-foot-tall gorilla becomes the pet and protector of an orphan girl, to the 1951 *Bedtime for Bonzo*, which starred U.S. president-to-be Ronald Reagan, apes were and continue to be profitable endeavors; the second major remake of *King Kong* appeared in theaters in 2005. Hollywood also produced a string of ape movie flops like *King Kong vs. Godzilla* and *The Beast that Killed Women*, a story of a murderous gorilla that tiptoes nightly into a nudist camp to prey on topless women.

It was the 1988 *Gorillas in the Mist*, however, starring Sigourney Weaver as Dian Fossey, the real-life pioneer of an up-close-and-personal style of gorilla research, that gave our primate cousins some form of redemption on the big screen. It was perhaps this film that stimulated public awareness of the plight of gorillas and their disappearing habitat, and initiated a deeper appreciation for the similarities between humans and gorillas.

Today everyone is familiar with chimpanzees. They fill our imagination from an early age, from Curious George to National Geographic programming. The more we discover about them, the more we become fascinated by the striking number of similarities with our own existence—their social relations, their behavior, their desires and expressions, their intellectual skills. Jane Goodall's statement that "the structure of the chimpanzee brain is startlingly similar to that of the human" may have forever changed how humans regard these animals.

Even a perfunctory visual examination of a human brain and a chimp's will yield several anatomical and striking similarities. This is perhaps not too surprising, considering our nearly identical DNA. It may seem peculiar, but chimps are even more closely related to us than they are to gorillas. Every day in the forests, they make decisions and cooperate as we do; they catch diseases as we do and make use of tools in a way that is similar to our use of them. They can learn the meaning of words and signs, master complex exercises, skills, and abstract concepts, and think in a logical fashion. They hug, kiss, and wrestle. They feel pain, fear, and joy. And they share something with us that we often consider uniquely human: they have a concept of self.

chapter two

TWO HISTORIES AFIELD

THERE WERE INFINITE constraints and innumer-
able logistical hurdles. There were unstable vessels
to work from and elusive animals that lived mostly
beneath the sea, swimming freely in an unknown
world. They were difficult to follow, and changing
conditions at sea could place the researchers themselves in peril with little or no warning. A dolphin

field researcher needed a variety of skills ranging from scientist to sailor. For these and other reasons, it is perhaps unsurprising that fieldwork at sea lagged behind the study of apes twenty or more years.

Although early data on the biology and population of large whales and stock management were collected in the commercial whaling community, the first glimpse of the social lives of small dolphins came from the tanks of marine studios. In the early 1940s, the comparative psychologist Donald Olding Hebb shifted his focus from primates to dolphins, collaborating on one of the first studies of dolphin social behavior in captivity. Following in his footsteps, scientists turned increasingly to the study of captive dolphins, who were trained primarily for public spectacles of synchronized leaps through hoops and ball play.

Seeking to escape the constraints of an aquarium tank and willing to brave hostile conditions and elusive animals, a few resourceful individuals looked seaward in the hope of understanding the lives of wild dolphins observed in their natural habitat. In the late 1950s and 1960s, David and

Melba Caldwell, a husband-and-wife team who had studied animals in captivity, contributing to the understanding of social relations of small cetaceans, ventured offshore in the Gulf of Mexico for what was probably one of the first opportunistic behavioral investigations of wild dolphins. John Prescott and Ken Norris, working at the now-defunct Marineland of the Pacific, were the first to discover a dolphin's ability to echolocate. Their early work on dolphins in the wild and that of the Caldwells helped to set the stage for other studies of these free-ranging, fast-moving, and still unfamiliar animals.

As cetacean scientists broadened their focus to include the open ocean, methodologies became more refined, and technologies were developed to solve the particular problems encountered in this kind of research. Inspiration for these methodologies and new technologies often came from the work of other animal behaviorists. For example, primate biologists had developed a method for recognizing and tracking individual mountain gorillas by collecting nose-print diagrams, and chimpanzee individuals were being identified by cataloging their unique facial characteristics. Following their lead, Bernd Würsig and other dolphin biologists created a system of photo-identification that utilized the natural nicks and marks on dolphin dorsal fins as a unique fingerprint to identify individual animals. The ability to recognize individuals over periods of years or decades launched a new era of long-term cetacean research and opened the door to a deeper understanding of the structure of dolphin societies.

The everyday presence of bottlenose dolphins near coastlines and their cosmopolitan distribution in temperate and tropical seas made them one of the most studied cetaceans worldwide. Randall Wells, a biologist at the Chicago Zoological Society and his collaborators at the Mote Marine Lab were among the most persistent followers of dolphins' flukeprints in the 1970s. Equipped with notebooks, cameras, and tape recorders, they spent over three decades observing a forty-square-mile area of Florida's Sarasota Bay from their small fleet of research vessels, running what is now recognized as the world's longest study of wild Atlantic bottlenose dolphins. In good and bad weather, they chased four generations of resident dolphins, photo-identifying over 2,500 dolphins and opening a window on the social organization of groups based on sex, age, and reproductive conditions. They established family trees, studied the reactions of pollutants in the water and the effects of interactions with humans, and set the groundwork for the conservation and management of small cetaceans.

A decade later in the waters of Shark Bay, Australia, Richard Connor, Rachel Smolker, and Janet Mann started a parallel project studying over four hundred identified Indian Ocean bottlenose dolphins. With its smooth, clear water, enviable weather, and dozens of animals to pursue for hours at a time, Shark Bay provided a paradise for dolphin research. While animals were bow riding upside-down, researchers were able to determine the sex of many local individuals, facilitating an understanding of their societies and the relationships between individuals.

Scientists in Shark Bay came to know their subject animals so well that they could recognize named individuals by sight. In the company of these dolphins, they discovered that the animals displayed associations similar to the ones found in Sarasota, but there was more. Male dolphins formed alliances with other males to cooperate, sometimes violently, against other alliances in competition for resources. This discovery revealed another face of dolphins; behind their eternal smile, these creatures were capable of violence not only toward other species but also toward their own.

In the late 1960s, Ken Norris chose Kealakekua Bay on the Big Island of Hawaii to initiate his research on the resident spinner dolphins. Over the next thirty years, he carried out four different studies and came to be known informally in cetologist circles as the grandfather of dolphin research. In their book *The Hawaiian Spinner Dolphin*, Norris, Wells, Würsig, and his wife Melany Würsig presented the first complete scientific natural history of a wild dolphin species ever written. One of their first discoveries was that spinner dolphins have an extreme sensitivity toward intruders, including researchers, which presented a new set of problems in observing them.

Norris wasn't an ordinary research scientist;

he used his extraordinary understanding of the natural world as well as old-fashioned ingenuity to study these difficult animals without actually interacting with them. He designed what became known as the SSSM or "semi-submersible seasick machine." Though constrained by lack of financial support, his first prototype was impressive, an airplane fuel tank with a welded-on observation cylinder. The field tests proved unsatisfying, however, and Norris began work on a second generation of underwater viewing vessel. The *Maka'ala,* which means "the watchful" in Hawaiian, lived a short life and ended up sinking in a rainstorm.

The third generation was another story altogether. The *Smyg Tittar'n,* Swedish for "the tip-toeing looker," eventually evolved into a functional and sophisticated eight-meter research vessel of 4.5 tons. An observer could climb down into a chamber with an underwater viewing window and record the dolphins' behavior. Despite being limited to inshore work due to its weight, it was flawlessly in line with Norris's approach of examining the dolphin world from the perspective of a dolphin. The *Smyg Tittar'n* provided a truly amazing window on dolphin underwater life . . . but the side-effect of turning human spectators green with sea-sickness was never solved.

Norris collected data by any means available, including data from land-based stations, aerial surveys, boat surveys, and observations from the *Smyg Tittar'n*. The length and breadth of this data brought to light the high degree of fluidity in dolphin school membership, the high level of cooperation among individuals, their tendency to use certain habitats for specific and repetitive purposes, and the importance and length of parental care and instruction in the development of offspring. Norris was also one of the first to use the human concept of "culture" when describing dolphin societies—a controversial assertion that now is gaining acceptance worldwide in the study of cetaceans.

With its distinctive tall fin, striking black and white markings, and a cosmopolitan range, another dolphin, the killer whale or orca, was a perfect candidate for long-term photo-identification studies. The world's authorities on these animals were Michael Bigg, Ken Balcomb, John Ford, and Graeme Ellis, whose studies in Washington state and British Columbia began in the 1970s. Their work focused on the complex social organization prevalent in this species, an organization that includes pods, matrilineal groups, and acoustic clans, as well as amazing foraging specializations and cooperative group hunting. Their investigations not only improved our knowledge of these whales, but helped to diminish their public reputation as ruthless killers.

Over three decades of long-term field studies have laid a healthy foundation for a growing number of students, volunteers, and researchers who are now continuing the pioneer research of their mentors, venturing into previously unstudied seas and looking at still-unfamiliar dolphin species. Although advancing technology has provided a vast array of aids that the first scientists could only dream of, like satellite tags, GPS, laptop computers, remote global imagery, and better foul weather gear, today's cetologists must still struggle with the perils of working at sea.

The Veil Lifts on the Apes

In the late nineteenth and early twentieth centuries, great apes were zoological curiosities, obtained at great expense by expeditions to the most remote parts of Africa. Gorillas were the most sought after, because of their imposing size. If they survived the sea voyage to London or New York, they rarely lived more than a few months. A visitor's guide to the Bronx Zoo from 1903 tells the visitor that the gorilla on display will likely live only a short time before succumbing to "its own sullenness." Keepers had little idea what their dietary needs were, let alone their psychological welfare behind the bars of antiquated zoological parks. They were treated as highly dangerous, even savage animals, although a caged gorilla is generally an apathetic, bored creature unlikely to harm others.

The public image of gorillas has evolved from savage beast to gentle giant. When the explorer and filmmaker Carl Akeley spent a few weeks trying to film mountain gorillas in the 1930s, it was the most intensive quality time anyone had ever spent with the enormous apes. Thirty years later, George Schaller, a famed wildlife biologist at the New York Zoological Society, spent a year living with mountain gorillas in the Democratic Republic of Congo. His account revealed gorillas to be timid creatures living in dire conditions, threatened by a surrounding sea of people. A year of field observation in those days was considered a Herculean feat, but twelve months limited the quantity and quality of information that could be gathered. Schaller used stealth to observe the gorillas, not the long, arduous process of accustoming the ape to one's presence.

Not until 1960 did the veil truly lift on the great apes. It was not an animal behaviorist but a fossil hunter who instigated the modern era of primate study. Louis Leakey, the son of English missionaries in Kenya, had established himself as an authority on fossils with the discovery of an early human skull, dubbed "Nutcracker Man," in Olduvai Gorge in 1959. He then turned to other ways to answer the great questions of human origins. Leakey reasoned that if one wanted to understand how early humans lived, what better way to do so than to study our nearest kin, who clearly resemble those ancestors, living in precisely the same sort of environment? He knew of a British colonial hunting reserve in the west of Tanzania, a place renowned for its stock of roan antelope, bushbuck, and buffalo. The place was called Gombe Stream Reserve, and it also had a healthy complement of chimpanzees.

Leakey needed an eager student to carry out his field study of chimpanzees. For this he turned to a young woman working as an assistant in his museum facility in Nairobi. Jane Goodall had gone out to East Africa from England as a tourist and secured a clerical position at Leakey's museum, hoping to pursue a life in science. Leakey gave her that chance. He sent her off to Gombe, thinking she might stay for a few months. Goodall immersed herself in the life of the forest and persisted through months of watching the rear ends of the apes fleeing her approach. Goodall's months turned into years, and as her subjects became accustomed to her presence, she painted a portrait of chimpanzee society that has given us a whole new view of what it means to be human. Hunting, meat-eating, and tool use were all traits thought to be uniquely human until Goodall's discoveries in the early 1960s. Then in the 1970s, Goodall and her growing legion of students and field assistants reported infanticide, cannibalism, and warfare—a dark side of the apes that reflected so well the dark side of ourselves. Over forty years after arriving at Gombe, Goodall still directs a project that is the longest-running study of wild animals ever conducted. Goodall herself is a leading voice for environmentalism and environmental education worldwide.

Nearly a decade after sending Jane Goodall to Gombe, Leakey gave another young woman an opportunity in a field dominated by men. Dian Fossey was an American occupational therapist who had gone on an African safari and came home determined to return to Africa to study wild animals. Leakey helped set up Fossey in the Virunga Mountains, a volcanic chain in east central Africa, where in the face of civil unrest, health problems, and numerous other hardships, she established and maintained a research station she named Karisoke for more than ten years. Fossey's triumph and tragedy are well known to many through *Gorillas in the Mist*, a Hollywood film based on her book. Fossey persevered through numerous tribulations, but she made the tragic mistake of trying to protect her gorillas at gunpoint rather than through community outreach and education of local villagers. This approach eventually resulted in her murder by a presumed poacher at her research station in 1984. Immediately following her death, a new generation of gorilla researchers and conservationists took over at Karisoke and focused on involving the people and government of Rwanda as protectors and beneficiaries of the gorillas. They established ecotourism as a major economic force in this poor, overcrowded nation, and though the horrors of war and genocide have taken a toll on the gorillas' future, the outlook is better today than it has been for some years.

Not long after Leakey sent Fossey off to Africa to study gorillas, he met Biruté Galdikas, a doctoral student at the University of California, Los Angeles, who expressed her desire to study orangutans. The result was a long, arduous study of the most inaccessible of the great apes. Galdikas reported previously unknown features of orangutan behavior such as male courtship calls and the occasional occurrence of rape. Because orangutans are generally solitary, it took years for Galdikas to compile the rich portrait of their societies that Fossey and Goodall could see in a much shorter time watching the interactions of entire social groups. Galdikas stayed on in Borneo for decades as Goodall had in Africa, spearheading the Indonesian conservation movement and butting heads with government officials time and again in efforts to stop the rampant illegal logging the government tolerates in its own national parks and reserves. Like Goodall, her current role is as a social activist, and like Goodall her energy is often focused on the plight of orphaned apes. Galdikas and a number of other conservationists in Indonesia have established a series of rehabilitation centers, intended to return orphaned orangutans to the wild after a period of re-education to the ways of wild apes. This is no easy process; it requires forests empty of apes (so the released orphans do not infect their wild cousins with diseases they may have acquired in captive life) and thousands of hours spent teaching a maturing orangutan all the skills it would have normally learned from its mother.

What we know about the last of the great ape species, the bonobo, is a fraction of what is known about each of the others. Unlike chimpanzees, which live in a variety of habitats across the equatorial belt of Africa, bonobos live only in a much smaller region of flat tropical forest in the center of the Congo Basin. The area is remote, accessible mainly by river and a few small dirt airstrips hacked from the rain forest. And it also lies in the heart of the war-torn Congo region, making the various field sites dangerous for researchers during times of war. At all times there is pressure from people poaching wildlife, including the bonobos. The bonobo was first described in the 1970s, when a team of Japanese scientists surveyed its remote forest world. But as early as the 1950s, European scientists who had observed bonobos in zoos had been writing about the "make love, not war" ape, as the bonobo came to be nicknamed. It was noted that they tend to have sex liberally in a variety of humanlike positions and contexts and show little preference for males or females in their choice of partners. Only in the 1980s did in-depth research begin, and it was hampered by the years needed to accustom shy wild apes to a human presence. A group of American researchers led by Randall Susman began a study at a site along the Lomako river in the Democratic Republic of Congo, then Zaïre. They saw many of the same behaviors that captive bonobos had displayed, though at much lower frequency than in the wild. The vagaries of grant funding and civil wars repeatedly interrupted their work, and after a decade there was still much to be learned about the apes. A German-led research team arrived in the same area later, followed by a Belgian team a few miles away, and through the 1980s and 1990s our knowledge of bonobos increased exponentially.

But much of what we know about bonobos is the result of work by Takayoshi Kano, a Japanese scientist from Osaka who established a research station at Wamba, also in the steamy jungles of Congo. Unlike the Americans at Lomako, Kano's team cleared a patch of rain forest and planted sugar cane, which they use as a massive candy store to attract the local bonobos. The result was apes who rapidly became accustomed to people, and therefore highly observable. This allowed Kano and associates to make more detailed observations of wild bonobo social behavior at close range than anyone else had done previously. Many of the sex-laden images of bonobo life as seen in captivity fall apart under scrutiny of their natural behavior in the rain forest.

Although it may seem that we have been watching great apes for a long time, the era of close observation and modern research methods is less than four decades old. It cannot last much longer. Several decades from now, the apes, if they exist in the wild at all, will be confined to tiny islands of forest in which their natural behavior will have been influenced by the many human disruptions that make their lives ever more difficult today. We will look back on the last forty years as a golden age for the apes; a time when their numbers were falling but still healthy. Their forest world had opened up, modern transportation had improved to allow researchers into their forest realms, and modern science had the means to study them in great detail. All this will almost certainly be gone during the lifetime of our children, if not within our own.

Even the depths of the oceans no longer represent a realm free of human presence as technology marches onward. The logistical difficulties encountered in the study of apes and cetaceans in the wild may very well be what protected them until now. What a loss it would be if we allow ourselves to pressure these incredible animals out of existence. Perhaps their salvation lies in an increased human understanding of their lives and behavior.

chapter three

SWIMMING WITH DOLPHINS,
SWINGING WITH APES

IF A SOCIETY is a group organized for some common purpose, then that implies the society has some knowledge of itself, perhaps even some degree of will involved in the society's formation and maintenance. Is this the case with dolphins? Some scientists describe social structure in animals as the patterning of relationships, but they may be ignoring the more subtle complexities of group organization.

As scientists we try to make sense out of the various behaviors we observe in the field. By limiting our observations to the measurement of patterns, are we overlooking a more complete picture of animal society? As observers of this complex fabric of life, we can't overlook the sheer beauty and balance of what we see unfolding before us. This, too, must be an integral part of an animal society. This, too, should figure into any definition. If we are to understand these societies, we probably need to revise our concept of intelligence and learn to think outside of our own species.

A Society under the Waves

All around me I see dolphins. I feel as though I could be part of a family, somehow different from my terrestrial one. At the bow of my boat, there's a group of exuberant bottlenose teenagers playing among themselves. Some carry strings of kelp on the end of their snouts while others frolic with small pieces of plastic or jellyfish. We are all following one of those "freeways" called fronts, which form when water masses of different temperatures meet in the ocean.

Most dolphins are social animals and, like apes and humans, derive more advantages than disadvantages from living in a group. Schools provide a dolphin with protection from predators, a food source, and a convenient place to meet fertile sexual partners.

Bottlenose dolphins are the most well-known and studied cetaceans. They spend their lives in what are known as fission-fusion societies. "Fission" means that the members of the local breeding population are continually splitting up and going their separate ways. And "fusion" means they always come back. A fission-fusion society may consist of several to many schools, the composition of which may be constantly changing on a daily or even hourly basis. The complexity of these societies, coupled with the difficulties of studying animals at sea, present quite a challenge for scientists bent on understanding these cosmopolitan animals.

Who is with whom? Who is *not* with whom?

These are important questions for dolphin researchers. Figuring out who, but also why, when, and how long different individuals stay together is very demanding due to the wide array of social strategies found in these animal populations. To illustrate how confusing this can be, it's enough to look at three populations of bottlenose dolphins— one swimming in Shark Bay, Australia, another in Sarasota Bay, Florida, and the third in waters of Moray Firth, Scotland. In Sarasota and Shark Bay, females in a school may be either relatively social or completely solitary, avoiding the presence of other individuals in their group. Males may travel alone or form strong relationships with other males, bonding in alliances comparable to those of their primate cousins. However, when we travel half the world away to Scotland, the females behave similarly but the males act quite differently— they form no alliances at all.

The unpredictable composition of a dolphin school becomes even messier when we look into the high variability in foraging strategies used by bottlenose dolphins at different locations. In some populations, they feed using a complex group strategy that involves ingeniously circling and herding fish schools while a few individuals at a time dive to catch prey. Several other animals stand guard nearby, occasionally switching jobs so that all members of the school may feed. In the Bahamas, things are different. Dolphins work alone by diving nose first into the sandy sea bottom to seek a yet-unidentified solitary prey, leaving the seafloor riddled with face prints that resemble small volcanoes. In Shark Bay, dolphins chase fish schools by swimming belly up near the surface. Prey are then either taken underwater or tossed into the air and snatched. In yet another technique prevalent on the coastlines of Georgia and South Carolina, bottlenose dolphins search out a fish school and, once located, cooperatively herd it to the muddy shoreline of tidal creeks, forcing the prey to jump out of the water onto the beach. When the fish are helplessly stranded, the dolphins charge the muddy beach, sometimes throwing their entire bodies out of the water to grab a fish, then leaping back to the safety of the water.

This diverse repertoire of foraging behaviors emphasizes the ability of these animals to capitalize on their own aptitudes and to adopt either solitary or group strategies for hunting, depending on a variety of factors including the availability and type of food sources and the surrounding environment. If we think for a moment about accomplishing any of these tasks and the many levels of communication, organization, innovation, and learning capacity that these strategies require, we begin to gain an appreciation for the level of intelligence inherent in a dolphin society.

Interspecific differences in foraging techniques are plentiful in cetacean field research observations. At one end of the spectrum there are some species of river dolphins like the susu, a nearly blind species on the brink of extinction in South Asia that somewhat resembles a pink aquatic anteater. The susu generally behaves as a solitary hunter, side-swimming along the bottom, endlessly nodding its head to scan for fish and obstacles with its sophisticated biosonar—a system used by dolphins to see and detect objects with sound. At the other end, we find huge schools of common dolphins—sometimes containing thousands of individuals—that can fan out to cover a vast amount of water in their search for patchily distributed fish schools in the open ocean. These cooperative groups may, later on, separate into smaller social units that disperse in different directions.

Somewhere in the middle of the spectrum are killer whales, cosmopolitan marine mammals living from the polar latitudes to the equatorial regions. Among the most distinctive members of the dolphin family, they also live in highly coordinated and complex societies based on communication and group cooperation. Studies in the northeastern Pacific show astonishing differences in dietary specializations between two sympatric populations of killer whales. These populations, referred to as resident and transient, live in the same coastal waters but are socially isolated from each other. They differ in morphology, genetic structure, distribution, and behavioral patterns. Resident groups feed on fish only, whereas transients feed mainly on marine mammals. Residents also live in long-term, large, and stable pods formed by several maternal lineages, whereas in transient populations all of the offspring—except for one male—disperse from their maternal pod. The offspring of tran-

sient whales, however, continue to live in their natal range, displaying what is called locational
philopatry. Transient whale pods of up to four individuals are much smaller than residents whose
numbers can reach two hundred in a pod.

No record of diet specialization like the one
found in the sympatric populations of killer whales
has ever been observed elsewhere in the mammal
world. These dietary strategies seem to have been
refined over a long period of time spanning multiple generations. In a clever and mutually exclusive
way, resident and transient whales have developed
their group foraging tactics to increase both rate
of prey encounter and prey capture success. Depending on the type of prey, resident and transient
whales have skillfully learned how to adopt different detection strategies, using echolocation when
looking for fish or passive listening during a hunt
for marine mammals.

Living in a school of common dolphins or killer
whales clearly requires much contact and strong
communication among individuals. In highly
multimodal animals like dolphins, this can be attained through visual, tactile, and acoustic means.

As in human societies, this voluminous exchange of information, transferred from one individual to another, forms the foundation for these coordinated social organizations.

In large or small schools, in fission-fusion societies or not, in coordinated foraging groups or as solitary hunters, cetaceans live in all oceans of the world. As diverse as their respective environments, they can be as different from each other as are the frozen seas of the arctic from the clear warm waters of the tropics. Different species may resemble each other physically, or their diversity may be obvious, as illustrated by the tremendous size difference between sperm whales, the largest at over eighteen meters in length, and the diminutive vaquita, an endangered species of porpoise indigenous to Mexico's Sea of Cortez, measuring just under a meter.

In their everyday life, all of these species must use context-dependent, complex social signals and draw from an extensive variety of strategies to resolve any given task, including not only foraging but also engaging in courtships, maintaining relationships and hierarchies, or warning other individuals of peril.

Like food, sex plays an all-important role in dolphin societies. Sex is essential both for reproduction and for communication. As observed with bonobos, sexual but nonreproductive behavior in marine mammal species like bottlenose dolphins and spinner dolphins is important in mediating social relationships. So-called "goosing" or rostrogenital contact is widespread in dolphin societies. This genital check-up, in which an individual rubs its beak into the genital area of another of the same or different sex, may tell a dolphin the reproductive state of the inspected animal. In the underwater world, it is neither unusual nor outrageous to find infant males trying to mount their mothers, young males sexually harassing older males, older males mounting calves, or an adult male mounting other males to express dominance.

Bottlenose dolphin males reach their sexual maturity between ten and twelve years of age; females between five and ten. As in many dolphin species, bottlenose males and females do not display sexual dimorphism, meaning that both sexes look pretty much the same to a human observer.

After a pregnancy of twelve months or so, a female dolphin gives birth to a single calf. The dolphin females invest a great deal in their offspring, as humans, great apes, and other mammals do. The birth of a dolphin calf marks the beginning of the strongest bond found within a school, that of a mother and her calf. They will be united for several years to come. At its mother's side, under her guidance and devotion, the calf will gradually be taught how to survive in the challenging ocean world. Understanding parental "love" in species other than ours is not easy, mostly because this kind of emotion is thought to be for humans only. But those of us who have observed the meticulous care of a mother for a calf, who have heard the pleading and calling of a dolphin mother suddenly separated from her offspring, or who have witnessed a mother lingering for hours near a lifeless calf, can't help but note the apparent similarities between dolphin and human behavior in such circumstances.

As they develop and grow older, the young dolphins will become increasingly more independent. After several years of living together, a mother and calf will go their separate ways. In bottlenose dolphin schools, some mothers will go off alone while others may stay in the school, forming "bands" with other mothers who are still caring for their calves. After reaching sexual maturity, males and females continue to breed well into their forties and live into their fifties.

In learning to evaluate and understand dolphin societies, we encounter a vast range of behaviors that include the long-term bonds between mothers and calves, alliances among males, group cooperation, complex feeding strategies, and more. But these diverse behaviors are not found only in dolphins. Surprisingly, these behaviors are also found in primates. A comparison of the societies of primates and dolphins reveals many parallels, despite the utterly different appearance and habitats of these two groups of mammals.

A Society in the Forest

Chimpanzees are the most versatile of the apes. They live across a wide swath of equatorial Africa, from Senegal in the west to Tanzania in the east, in habitats ranging from dense tropical rain forest to open, nearly treeless grassland. In the center of Africa, vast expanses of forest still are home to large populations of chimpanzees. But these forest tracts are no longer safe havens. Logging ventures are tunneling their way into the carpet of greenery, carving up ape habitat.

Most people are familiar with chimpanzees, but a close cousin of the chimpanzee has only more recently come to public attention. The bonobo looks very much like a chimpanzee. Early reports referred to it as the pygmy chimpanzee, but the two species are nearly the same in size, color, and every other physical characteristic. As we shall see later, it is in behavior that the two apes diverge.

Bonobos no doubt once were chimpanzees. Their geographic distribution in Africa is below the vast bend of the Congo River. On the northern shore are chimpanzees, and also gorillas. The tracts of tropical rain forest below the arc of the river belong to bonobos, although their total distribution is a tiny fraction of that of the other African apes. Unlike the environmentally catholic chimpanzee, bonobos are limited to lowland forest. Within their range they number no more than a few tens of thousands, and perhaps far less; ongoing political chaos in the Democratic Republic of Congo has prevented any recent census-taking.

The third of the African apes is the largest of all primates—the gorilla. Gorillas come in several varieties. In most of Africa, western lowland gorillas, the kind found in zoos, are the norm. These are majestic animals with silver-tipped, gray-black short hair and chestnut caps. They haunt the flat, rain-forested regions of western and central Africa. Lowland gorillas number in the tens of thousands, though numbers are believed to be falling fast. Farther east, a blacker form, the eastern lowland gorilla, lives in both low-lying and mountainous forest. Eastern lowlanders are increasingly rare, their numbers dropping rapidly in the wake of civil war and much poaching in the limited forest areas in which they live.

The best-known variety of gorilla is also the rarest. Mountain gorillas live in two separate, nearby forest kingdoms. Along the spine of the Virunga Volcanoes, a chain of cone-shaped volcanic mountains in east-central Africa, live 350 mountain gorillas. Straddling the border between the troubled nations of Rwanda and the Democratic Republic of Congo, and dipping just a bit into Uganda as well, the Virungas harbor the animals made famous by the late Dian Fossey.

Less well-known than the Virungas mountain gorillas are those living just a few miles away in a nearby range of rugged hills, Bwindi Impenetrable National Park. Bwindi is home to about 300 more mountain gorillas, distinguishable from their montane cousins only by a few superficial features. Between Bwindi and the Virungas, there are 650 or so mountain gorillas left on Earth. Their habitat is protected, but their numbers are precariously vulnerable to epidemics, civil war, poaching, and other social ills.

In the forests of Asia live the last of the four great apes, and the most enigmatic, poorly understood ones, too. Orangutans are shaggy-haired, red apes that live in the dense tropical forests of the Indonesian islands of Borneo and Sumatra. The plight of the orangutan may be worse than that of all the other apes, or at least it is the best documented. Rapid and often illegal cutting of the Indonesian forests that are their homes are leaving orangutans without any place to live. This is happening at an accelerating rate despite the best efforts of conservationists to stem the tide. Before we have learned the inner details of the lives of these, our fourth nearest kin, they may well be gone.

Chimpanzee and bonobo societies are superficially quite similar. Both species live in fission-fusion communities and spend their days searching for fruit with an occasional morsel of meat. But in their gender relations they have gone their own ways. Chimpanzees live in male-dominated societies in which males solicit sex, sometimes brutalizing females who do not produce. This is not to say that females are always submissive; but if they do not respond to a male's advances, males may attack them violently. As a male matures, he rises in dominance rank, passing each female one by one until he begins to challenge the lowest-ranking males. Such is life for female chimpanzees; even the top-ranking female is subordinate to the lowliest male.

Bonobo females avoid such maltreatment by the male gender by bonding together, forming co-alitions that ward off male aggression. These alliances sometimes even allow females to dominate males in some contexts. Bonobos have become famous as "feminist" apes, but this female tactic is really all about living with other females more than harassing males. By establishing a way to co-exist peacefully with one another, female bonobos avoid the sort of squabbling over food that prevents female chimps from forming alliances.

Gorillas, meanwhile, live in far more conventional arrangements. Walking through a forest in Uganda in the company of a mountain gorilla group is like being out on a family stroll in the park. They walk slowly, kids exploring every little nook and cranny as we go, moms attentive, and a silverback male looking worried. Weighing in at upward of 400 pounds, gorillas are among the brainiest land animals ever to live on Earth. They are the most ground-based of the apes—their enormous bulk doesn't suit them well for treetop acrobatics, though they sometimes try—and also the most purely vegetarian. Whereas chimpanzees and bonobos love meat and sometimes go to great lengths to get a scrap, gorillas disdain even the insects that chimpanzees relish. They are consummate browsers, eating foliage and fruit in tropical forests across Africa. On the surface, gorilla society seems orderly and staid compared with the high-energy comings and goings of chimpanzees. But underlying their majestic aura, gorillas are a bit edgy themselves. As we push through dense undergrowth, the silverback suddenly stops and begins to make anxious grunting sounds. The entire group freezes. After long minutes of silent tension, the cause of the silverback's worry appears. Across a small clearing about fifty yards away, another silverback's head pops through the wall of greenery. There is a face-off—the two grunt and roar and pace. The females and their young retreat to the thickets to let the two males stand down. After twenty minutes, the males back down and move apart, and the encounter ends.

It all seems rather ritualized and utterly non-physical, but important life decisions may be made during such encounters. Females in gorilla society tend to migrate from group to group, thwarting male attempts to monopolize. They use their feet to exercise their choice of mates. Such departures are likeliest to happen in the time after encounters between two rival groups. This suggests that female gorillas are using the stand-offs between silverbacks to size up both their own male and the one challenging him, and if they decide their own mate is something of a loser, they leave him. For all their bulk and muscle and bravado, silverback gorillas live in fear of their females walking out one day, and this motivates their solicitous protection of their group.

Compared to the other living apes, the orangutans are quite bizarre, both in form and behavior. Male orangutans of the Bornean variety possess enormous flanges of flesh around their face and a pendulous sac of flesh on the throat. The throat sac helps to produce mating and dominance calls, and the flanges are presumably attractive to females as well. Males are enormously larger than females, nearly twice their size and bulk.

In behavior, the orangutan's uniqueness truly shines. Unlike the highly sociable gorillas, chimpanzees, and bonobos, orangutans relish their solitude. They are not only the most solitary of the apes, but among the most solitary of the higher primates. A female will defend her territory against all comers, while an adult male will try to maintain a sexual monopoly over several females. This is a good strategy on the male's part, but it ultimately fails because adult male orangutans are not only big but also slow. They cannot cover their vast forest range quickly. This allows other males who don't have their own coterie of females to maraud.

All the great apes share a few traits. They are among the largest of the primates. They have life cycles not unlike our own: an eight-month pregnancy, a several-year period of infancy and childhood, puberty in the early teen years, and reproduction from midteen years through their forties. And they are the brainiest creatures on Earth, along with the dolphins, whales, and ourselves. The parallels do not end there; they include the best and the worst of human nature, from warring to loving.

Hunters

Dolphins have no hands to make tools, yet they are efficient predators who use both their agility and their braininess to achieve hunting success. They are original, intelligent, and cooperative in their hunt for a meal.

Among all dolphins, transient killer whales triumph for their cooperative hunting. As the top predators of the oceans, they reveal breathtakingly complex tactics when attacking other mammals. Transient killer whales are the only cetaceans that attack and eat other marine mammals. They may prey on small marine mammals like seals lying on a beach, but they can also single out and strike whales much larger than themselves without any sign of fear or hesitation and with a high degree of predatory success.

For example, a few miles off the Oregon coast, a pod of transient killer whales readies itself for an attack. They have found a female gray whale and her calf traveling steadily northward on their migratory route from breeding grounds in Baja California to the cold feeding grounds of Alaska. The killer whales trail a short distance behind, pushing the gray whale and her calf forward with their presence, waiting for the calf to tire and the mother to slow her pace. The killer whales do not hurry; their powerful bodies are well adapted to this, allowing them to swim for long distances without any sign of fatigue. Not so, however, for the newborn gray whale. Fleeing the attack has exhausted the calf, and the mother must slow down at some point to allow her offspring to rest. This is the moment for which the killer whales wait patiently, this the opportunity that they have engineered.

In a synchronized action, they launch an assault on the now almost motionless calf. The gray whale mother tries to protect her calf, fending off lunges from multiple directions with her powerful tail, though her efforts are eventually in vain. The killer whales attack from all directions in relentless synchrony, tearing off chunks of flesh and biting the calf on the flukes to disable any last attempt at flight. Successful, the killer whale pod now swarms around the dead baby whale, feeding at will until their hunger is sated and they depart *en masse*. Distraught over the death of her calf and tired from her fight and her long journey, the gray whale mother lingers beside her dead offspring before resuming her journey northward. Only through complex communication has the pod of killer whales managed to feed themselves—a single attacker would stand no chance.

What we know about dolphins as "hunting machines" comes mostly from our surface observations. Standing on the deck of a research ship on the high seas, sitting near shore in a tiny inflatable, or looking down from a cliff with a pair of binoculars, we attempt to glean what we can of dolphin life from their appearances at the surface. But the three-dimensional environment beneath the sea still holds the secrets of how these hunters exploit the water column for their foraging activities. Although we humans must use all sorts of complicated equipment to explore the underwater world, for a dolphin, diving to a depth of fifty meters or more for a meal is simply a daily routine. To make the most of underwater food sources, dolphins have evolved sophisticated diving adaptations that allow them to spend time at depth while keeping their energetic expenses to a minimum. Suspending respiration, bradycardia (a low heart rate), and peripheral vasoconstriction are some of the physiological tools employed by these diving champions.

Being such accomplished ocean hunters make dolphins a valuable asset for other ocean dwellers in search of a meal. In my work in the coastal waters of Los Angeles, I frequently observed sea lions in proximity to common dolphins during feeding and foraging activities. Here were two species, traveling and feeding together, with no evident hostility or competition. To me, this seemed unusual and warranted further investigation. What I discovered was that the sea lions were capitalizing on the superior food-finding ability of echolocating common dolphins to find their own prey. In some cases, keeping up with traveling dolphins presented a challenge for sea lions. Dolphins can swim much faster and more efficiently. By videotaping sea lions in their pursuit of dolphins, we were able to observe pinnipeds spyhopping to get a better fix on the dolphins' movements, adjusting their routes and speed accordingly. Once at the feeding ground, dolphins and sea lions shared meals with no rivalry. Then each went its own way.

On first evaluation, the sea lions seem to gain all the advantage from these aggregations, but dolphins may also profit from this sharing of resources. For instance, numerous predators in a feeding area rich in prey can scare fish into a protective schooling behavior, thereby increasing the potential catch for sea lions and dolphins alike.

Although dolphins are expert predators, using their intelligence to find widely scattered schools of fish or other animals and their speed and agility to catch them, at least one great ape, the chimpanzee, also catches and eats animal protein. Chimpanzees hunt for monkeys, wild pigs, antelope, and a number of other small mammals. Few people believed it when Jane Goodall first reported meat-eating wild chimpanzees, but hunting is a completely natural, routine part of chimpanzee life.

Most species of primate eat a bit of animal

protein, usually in the form of insects. Only a few of the higher primates eat other mammals. In the tropical forest of Central and South America, capuchin monkeys dine on a variety of smaller animals, including squirrels and other rodents. Baboons also hunt small mammals such as rabbits and antelope fawns. But only chimpanzees engage in the sort of systematic predatory behavior that we believe was central to the lives of our earliest human ancestors. In all forests in which they have been studied intensively, they prey on a variety of vertebrates, including dozens of kinds of other mammals. The red colobus monkey is the most frequently eaten prey. In some years, chimpanzees in Gombe National Park, Tanzania, kill more than 800 kilograms of prey, most of it red colobus.

Hunting is social, whether by dolphins for fish or by chimpanzees for monkeys. Chimpanzees, usually adult males, attack groups of monkeys they encounter during their rambles in search of fruit in African forests. The chase, capture, and kill are heart-stopping, often gruesome, and always illustrative of the chimpanzees' social nature. To a lion the zebra it is chasing may be only meat, but to a chimpanzee the chance to kill and then share prey is not only nutritional but socially significant as well. The monkeys, pigs, and antelopes the chimpanzees capture sometimes become pawns in the social dynamics of the group—currency to be used to negotiate new alliances, rub salt in the wounds of old rivals, and secure status that a chimpanzee without prey cannot.

A chimpanzee hunt for monkey prey begins with a chance meeting in the forest. Chimpanzees travel from valley to valley in search of ripe fruit, and from time to time they walk under a tree in which a group of colobus monkeys sits. Colobus are long-tailed, treetop-dwelling monkeys that weigh perhaps twenty pounds—quite small compared to an adult male chimpanzee's over 100-pound bulk. Unlike most other types of monkeys, they don't usually flee when they spot chimpanzees approaching. Instead, they continue eating leaves until the apes reach their tree, and then erupt in a frenzy of high-pitched alarm calls and nervous looks. Mothers retrieve their babies and males cluster together in a protective shield to keep the chimpanzees at bay. At this point the chimps decide whether to continue walking or to take a shot at the monkeys. If they decide to hunt—and the mental factors in making their decision are as fascinating as they are unclear—the chimps begin to climb the tree. The male monkeys often descend to the first large fork in the trunk, trying to keep the chimpanzees out of the crown the way a metal baffle on a bird feeder is supposed to keep the squirrels at bay.

If the chimpanzees persist and get past the legion of male monkeys, several things can happen, all of them bad for the colobus. The chimps may rush toward a female carrying a baby, trying to take the baby as their prey, and the mother too if she fights back. More often she will almost passively allow the hunters to pluck the infant from her chest and kill it with a quick bite in the skull. The hunters may also capture and kill one or more male monkeys. This is the norm in some forests in western Africa, perhaps because chimpanzees there are bigger, and colobus males smaller, than in eastern Africa. These male monkeys are killed by being thrashed against a tree trunk or against the hard ground. The chimpanzees may continue to hunt after making one kill until two, three, four, or on rare occasions as many as a dozen monkeys lie dead. In cases like this, an entire group of monkeys may be decimated within minutes.

Forty years after Dr. Goodall first witnessed the hunting behavior of chimpanzees, we know a great deal about their predatory patterns. At Gombe, red colobus account for more than 80 percent of the prey eaten. Here, chimpanzees do not select the colobus they will kill randomly; most colobus killed are immature and often infants. In Gombe, chimpanzees are largely fruit eaters. Only 3 percent of the time they spend eating is spent consuming meat—less than in nearly all human societies. Adult and adolescent males do most of the hunting, making about 90 percent of the kills recorded at Gombe. Females also hunt, but more often receive a share of meat from the male who either captured the meat or stole it from the captor. Although lone chimpanzees, both male and female, sometimes hunt by themselves, most hunts are social.

The quantity of meat eaten is substantial. In some years the forty-five chimpanzees of the Kasakela community at Gombe kill and consume hundreds of kilograms of prey. In the dry season, the meat intake is about 65 grams of meat per day for each adult chimpanzee, not far below the meat intake by the members of some human foraging societies.

Jane Goodall had observed early on that her chimpanzees went on "hunting crazes," during which they would hunt and kill large numbers of monkeys and other prey. The most intense hunting binge we have seen occurred in the dry season of 1990. From late June through early September, a period of sixty-eight days, the chimpanzees were observed to kill over seventy colobus monkeys in forty-seven hunts.

Hunting by wild chimpanzees appears to have both a nutritional and social basis. Early researchers had said that hunting by chimpanzees might be a form of social display, in which a male chimpanzee tries to show his prowess to other members of the community. Richard Wrangham conducted the first systematic study of chimpanzee behavioral ecology at Gombe and concluded that predation by chimpanzees was nutritionally based, but that some aspects of the behavior were not well explained by nutritional needs alone. Researchers in the Mahale Mountains project reported that the alpha there, Ntologi, used captured meat as a political tool to withhold from rivals and dole out to allies.

It initially seemed to me that hunting must serve a nutritional function above all. Meat from monkeys and other prey would be a package of protein, fat, and calories hard to equal from any plant food. I therefore examined the relationship between the odds of success and the amount of meat available with different numbers of hunters in relation to each hunter's expected payoff in meat obtained. That is, when is the time, energy, and risk (the costs) involved in hunting worth the potential benefits, and therefore when should a chimp decide to join or not join a hunting party? And how does it compare to the costs and benefits of foraging for plant foods? The results were surprising. I expected that as the number of hunters increased, the amount of meat available for each hunter would also increase. This would have explained the social nature of hunting by Gombe chimpanzees. If the amount of meat available per hunter declined with increasing hunting-party size (because each hunter got smaller portions as party size increased), then it would be a better investment of time and energy to hunt alone rather than join a party. The hunting success rates of lone hunters are only about 30 percent, whereas that of parties with ten or more hunters is nearly 100 percent. As it turned out, there was no relationship, either positive or negative, between the number of hunters and the amount of meat available per capita. This may be because even though the likelihood of success increased with more hunters in the party, the most frequently caught prey animal is a one-kilogram baby colobus monkey. Whether shared among four hunters or fourteen, such a small package of meat does not provide anyone with much food.

Once the killing has ended, the consumption and ritual sharing begins. Sometimes a captor takes his prize behind dense foliage to have a private, well-earned meal. More often, the kill is eaten publicly, since the sharing or refusal to share appear to have social significance in chimpanzee society. A male who did not catch a monkey himself, but whose high status allows him to abscond with a carcass from a lower-ranking hunter, may simply grab a monkey away from his mate with impunity. The possessor of meat becomes a group focal point, and is subjected to intense begging. With hands outstretched, sometimes even inserted into the mouth of the meat-eater, other members of the group beseech the captor for a scrap of the rare treat of protein and fat. Allies get favors of meat, as do kin. Sometimes those who helped make the capture but didn't end up with the monkey also get a seat at the table. Sharing meat is a Machiavellian act with a clear, public message. Not only will a male avoid sharing with a rival, he may purposefully snub him in order to show everyone else present just how impressive he is and how unimportant the rival is. Anthropologists have analyzed chimpanzee hunting behavior again and again because it is thought that meat-eating made us human. In the dark days 2 to 3 million years ago, scraps of meat would have been very precious commodities worth hunting for, scavenging for, and then treating with the highest regard once a capture had been made. Tool use no doubt evolved because it enabled early humans to get more and better food then they could without tools. Eventually the little monkey and piglet prey were abandoned for big-bodied animals like zebra and wildebeest, which require cooperation to catch and sharp tools to carve up and steal away afterward.

The diverse hunting strategies employed by dolphin and ape societies are an excellent reflection of their social complexity. It takes agility and weaponry for a lone predator to hunt and kill its prey, but for a social animal, it takes far more than that. It takes intelligence and the ability to coordinate actions with group mates. But the social complexity of large-brained mammals cannot be determined by a description of hunting behavior alone. Digging deeper, we will examine in the following chapters the nature of the underlying relationships and ties that make this complexity possible.

chapter four

DOLPHIN AND APE SOCIETIES—
WHYS AND WHEREFORES

AT DAWN IN Gombe National Park, Tanzania, an orange sun is rising in the east, but it will be hours before it appears over the mountain ridge above us to shed warming light on the forest below. A party of chimpanzees is waking up. They roll over, look up at the morning sky, and slowly revive themselves on this chilly morning. One by one, they pull themselves upright in their nests. Each tree has a sleeping ape or two, and one towering *Chrysophyllum* tree holds several fresh nests and many old, wilted ones. Each chimpanzee sits sleepily on the branch supporting his or her nest, peeing onto the ground many meters below. Minutes pass and the forest grows light; shapes emerge out of the gloom that reveal themselves as chimps. They've silently descended from their sleeping trees and are sitting like boulders on the hillside.

At 5:45 AM, as if on cue, one of the older males gets up and begins to walk away from the sleeping area, heading north. Several other males immediately follow, while two males stay behind. Minutes later, these males turn and walk west, down toward the lake edge. A mother and her infant head off southward, alone. And a couple of young males just stay in the nesting area; later they will head east, up into the rugged hills. What started out at dawn as a nesting party of twenty-six chimpanzees has fragmented into at least five separate parties composed of one to eight chimpanzees each.

A little over seven hours earlier on the opposite side of the world, a frail rising sun is just beginning to light up the village of Rio Lagartos on the northern coast of the Yucatan Peninsula, Mexico. It is almost 6:30 AM at the end of the fisherman's rickety wharf. The dolphins are rarely late. Gordo, a chubby male bottlenose boasting a clear, deep notch halfway down his dorsal fin, is the first to appear in the morning mist. He makes his way slowly westward along the shoreline, followed by the rest of the gang, a football field behind. As the sun starts to brighten, their dark grey bodies pass in front of the pier in an organized procession. There are fourteen in total: a female with her calf and a dozen others. Twenty or so meters past the wharf, the procession clusters together in a circle next to a colorful parade of moored *pangas,* small skiffs used for fishing. Some are diving, others mill about at the surface. A few at a time, they explore the sandy bottom unhurriedly while another group of dolphins leisurely joins them from the opposite direction. They are now twenty-three in number, with a couple more calves swimming next to their mothers, all tightly grouped together in this uninspiring patch of murky water that likely hides a fishy meal. The dolphin circle suddenly unwinds in two lively threads. Five animals move steadily back toward the wharf in a formal procession; the others disappear quickly to the west. The sun is already high on the horizon, and what seemed for a moment to be a singular and cohesive group has reshuffled itself and divided once more. Handing over my binoculars to my assistant, I give up my spot on the wharf and head toward the village for a breakfast of *huevos con chorizo*.

Strong Bonds, Weak Bonds

In their social grouping patterns, dolphins and apes show some striking parallels. Chimpanzees and their cousins the bonobos do not live in stable groups as traditionally defined. Instead, they exist in a fluid, flexible societal arrangement that primatologists call fission-fusion polygyny. Polygyny means there are multiple males and females in the group. The arrangement is so lacking in apparent organization that it took nearly two decades from the start of Jane Goodall's research at Gombe to determine that there is order at all in chimpanzee society.

The first primatologists were convinced there was no structure to chimpanzee society, that life was just a sexual free-for-all in which females mated promiscuously whenever they were swollen and avoided males the rest of the time. But chimpanzee life is far more orderly and proceeds by a number of rules. Female chimpanzees are strategists who mate with males in hopes of securing their support, and more important their docility, toward their babies in future years. Males who have mated with a female and accept the ensuing birth as their own progeny are thought to be less likely to be aggressive to the female or her baby. At the same time, by inciting males to fight over her at the sight of an appealing pink swelling, a female may be able to judge which males are "keepers" and which losers in genetic strength and health.

Overarching this pattern of coming together and parting over food and sex, chimpanzee society adds another layer. All the chimpanzees in a local breeding group live in a community; a stable and long-lasting aggregation of males, females, and babies. Because males stay in the area of their birth while females emigrate at puberty, communities tend to be made of related males and less-related females. At about twenty to 120 individuals, these communities have well-defined territorial boundaries, the accidental crossing of which can lead to bloodshed.

Why such a complicated system? We believe the chimpanzee tendency to fission and fuse is related, as it is in dolphins, to both food and sex. The goal of a female chimp's life is to rear her babies successfully, and to do this she needs optimal access to good food. This means avoiding the competition as far as it is possible. So she spends much of her life searching for trees laden with ripe fruit, usually alone or in the company of her offspring. Males, meanwhile, have no such food and baby agenda. Their agenda is to find females and do what it takes to impregnate them. But female chimpanzees show an interest in mating only for about seven to ten days out of a thirty-seven-day menstrual cycle. During this fertile period the female carries a billboard of sexual availability; a fluid-filled pink sexual swelling that hangs from her rump. When her swelling is engorged, she seeks out males, and males seek her out, too. This leads to the formation of large, temporary parties of male and female chimps traveling the forest, the males vying for female affection and sometimes fighting over it.

Researchers have struggled to understand exactly what molds chimpanzee society. Certainly the dispersion of fruit across the forest has a lot to do with it. Leaves are widely available but low in nutrient content. Unlike leaves, which are available on every tree, fruit is only seasonally available, and its distribution is also patchy. A chimpanzee searching for fruit might need to walk several kilometers between trees bearing ripe fruit, and could arrive to find the next tree already picked over by other primates, birds, fruit bats, or a host of other animals. But the reward of a high-carbohydrate meal is enough to keep chimpanzees commuting among stands of fruit trees. Chimpanzees lead a high-energy life compared to many other animals, and their reliance on fruit—two-thirds to three-quarters of their diet—is both a cause and a consequence of it.

Their unconventional social system may be an adaptation to a life dependent on ripe fruit. Since fruit occurs in clumps—each clump being a tree—competition for food is likely among the members of a chimpanzee community. Females avoid such competition by foraging alone, carrying their dependent offspring. With females scattered across the entire territory of the community (although each female has a preferred area within the range), males attempt to maintain access to females while guarding their territory. When females emigrate at puberty or after, they may do so to avoid inbreeding with their fathers, who have remained in the community for life.

The community is therefore composed largely of immigrant, unrelated females and the cohort of males who have been born and have grown up in that same area. The only time during a female's life when she is highly sociable is when she is in estrus. Her sexual swelling becomes a magnet for males, and she herself is drawn to them. During this time of her five-week menstrual cycle, large parties of males and females and accompanying juveniles may form around such swollen females.

When highly desirable fruit trees are ripe at the same time that highly desired females are swollen, the community takes on an atmosphere of pandemonium. Males jostle over the females, fights break out among the male hierarchy, and everyone wants access to the same tree.

Although the influence of food on grouping is indisputable, there is also a clear role of social influences on grouping. Jane Goodall first stated that chimpanzee society was organized more around male access to females than around access to food. Researchers since have split on the question, and many have come to recognize the influences of both sex and food. For instance, among the famed Gombe chimpanzees in Tanzania, the presence of a swollen female in the community is a predictor of large party size. When a female is swollen, males tend to socialize around her, and party size increases. Large parties hunt more and are more often successful, engage in territorial aggression more readily, and exhibit higher levels of intra-community squabbling and competition for sex. In others words, stuff happens in large parties. Whether food or reproductively active females exert the more important influence on party size is unknown. It appears to depend on which community of chimpanzees one is watching.

Chimpanzee society is remarkably uniform across Africa. Everywhere chimpanzees have been studied, they live in highly fluid communities—never cohesive groups—made up of immigrant females and lifelong resident males. Some communities are large—one such at Ngogo in Uganda is over 120 members—whereas others are as small as twenty-five. Some communities are more cohesive than others. In Taï National Park in Ivory Coast, females appear to be more closely integrated into the community. In Gombe National Park, Tanzania, females are almost free agents, traveling independently, with males of the community mapping themselves onto the ranges of the females.

This form of society is mirrored in that of the closest kin of the chimpanzee, the bonobo. The bonobo also lives in communities centered around males from which females emigrate at maturity. The key difference between the two ape species is the conduct of females. Whereas female chimpanzees share a community membership without a great deal of interaction—and sometimes outright hostility—female bonobos have learned to get along. Reports coming back from central Africa in the 1980s stunned scientists who had watched chimpanzees; bonobo females seem to form coalitions bent on preventing males from exerting power. These alliances create a female power base in bonobo society that is largely missing among chimpanzees. Bonobos have been hyped in recent years as the "make love not war" apes for their alleged extreme sexuality. These claims have been met with some skepticism by researchers who have actually watched wild bonobos. But the female power sharing in bonobo society is indisputable, and has profound implications for using great apes as models of human evolution. Chimpanzees live in societies in which males wield power over females, sometimes brutally. Females exert themselves too, actively (if furtively) seeking out preferred mates and risking physical abuse from males if caught. But among bonobos, females are power brokers; in the enforced proximity of zoo enclosures, female bonobos have been known to attack and badly injure males, something almost unheard of among chimpanzees.

Their social organization may be fluid and

flexible, but other bonds among chimpanzees and bonobos are rock solid. The central bond is between mother and child. The life history of a female chimpanzee is not very different from that of a woman. The infant female chimpanzee is utterly dependent on her mother until about age four. She is then weaned off mother's milk but remains psychologically dependent on mom for several more years. By age ten or eleven she is pre-adolescent, and by about twelve she has experienced her first swelling. The sudden appearance of sexual advertisement in the maturing female causes males to treat her as a potential mating partner, and this dramatically alters her life in the community. At about this time she may begin to make temporary trips outside the community—visits, as it were— to neighboring communities. These visits will become routine and longer in duration. Sometime between age twelve and fifteen she will settle permanently in the other community and take up a new life as a breeding female and mother (although at some sites some females remain in their home community for life).

Although the female has been mating with

males since her early teen years, she will likely not become pregnant until about age fifteen, and will produce her first baby at age sixteen or seventeen. Her pregnancies will last about eight and a half months. From then on, she will give birth to one offspring about every four to five years until the end of her life, which can be nearly fifty in the wild (though typically disease and predators end it much earlier) and up to the sixties in captivity. The only striking difference between chimpanzee and human is menopause, which in humans signifies a fairly abrupt end to the reproductive lifespan, but after which the woman may live decades. Chimpanzees do not exhibit menopause of the human kind; a female's fertility may decline in her later years, but her reproductive life is usually brought to an end only when she dies. Anthropologists have long pondered this human-ape evolutionary distinction. The leading theory to explain human menopause is that there is an adaptive benefit to a woman who ceases her own reproductive output and turns to helping her offspring continue theirs. That is, she takes on the role of a grandmother.

Male chimpanzees, on the other hand, spend

their entire lives in the same place, in the company of their brothers, cousins, and other male kin. Males have the same long, slow life history that females have, except that as they enter their teenage years they tend to become less social for a time, traveling alone and living as wannabe adult males that do not find easy acceptance among the adult cohort. At some point in the mid-teen years, males begin to emulate their elders and join foraging parties. They attempt to hunt with the older males, although in the event that they catch a monkey or pig, it will be quickly repossessed by the larger hunters. On territorial patrols they are eager, if naïve, participants.

Over the ensuing years, males reach sexual and social maturity and begin to attempt an ascent up the dominance hierarchy. The chimpanzee dominance hierarchy is a central part of their society. It is not a simple ladder but a multi-faceted, complex set of dominant and subordinate relationships among males and females. As we already know, one universal about chimpanzee dominance is that even the lowest-ranking male is dominant to the highest-ranking female. So when young males want to achieve a level of status among the male hierarchy, they begin by systematically challenging each female in turn. Bluff attacks—rarely involving physical violence and injury—usually suffice to place a male in a socially superior position to each female. Once the highest-ranking female has been challenged and bested, he takes on the other males. But male dominance is highly dependent on context, and on one's allies. Instead of a linear pecking order, it is a sophisticated affair in which a male's status may flip-flop depending on which of his allies or rivals happens to be sitting near him at a given moment. Dominance contests are won not because of brute strength, but because of political cunning. Just as a president doesn't derive his strength from being six feet five inches, an alpha-ranked male chimpanzee can be small or large, so long as he knows how to manipulate those around him.

As a young adult male chimpanzee seeks the next steps in his rise in the male hierarchy, it is entirely insufficient for him to simply challenge each male above him. He needs to remember who is his ally, who is his rival, and the entire network of alliances swirling around him. Like a character in Shakespeare's *King Lear,* a male chimpanzee must not challenge a higher-ranking male when that male is accompanied by his best allies. Instead, he waits to find that male alone, or in the company of his rivals, and employs a "the enemy of my enemy is my friend" tactic. By picking his time and place carefully, he can defeat a male with the aid of other low-ranking males, even if none of them could consider challenging that male alone. Defeating the alpha doesn't have to mean physical violence, though it often does. It means demonstrating to him that the power base in the community has shifted and that, without power, the alpha needs to step down and get out of the way. In practice, most alphas who lose their status spend some time, often many months, nearly alone and return to the fold of the community duly chastened and much reduced in machismo.

In the best of all worlds, a maturing male is able to rise in the dominance hierarchy until he is ready to challenge the alpha. But most males plateau somewhere below this status, and end up using their alliances with both higher- and lower-ranking males to further their own interests in the community. There are rare exceptions. Goblin, of the Kasakela community at Gombe, began to challenge the top-ranking males when he was a ten-year-old juvenile, an unheard-of occurrence in chimpanzee society. It worked, too; by the time he was thirteen, an age at which most male chimpanzees are still trying to figure out how to live peacefully among adult males without inviting too much aggression, Goblin had toppled the alpha and become the top-ranking chimpanzee in the community. He held this status on and off for nearly fifteen years, and even long after he had passed his prime, he remained a "king-maker," flip-flopping his allegiances on a daily basis when it suited him. The long-term alpha male of the Mahale Mountains, Ntilogi, was an even more extreme example of chimpanzee overachievement. He was alpha for almost ten years before being overthrown. He then did the unexpected—after a time away from the community in exile, he suddenly returned and successfully took back his rank, which he held until his death at almost age fifty.

The careers of both male and female chimpanzees are therefore not so different from our own, except for the greater degree of gender balance our own species still strives to achieve. In the case of both males and females, there is a Darwinian bottom line: reproductive success. Males and females, whether dolphins or apes, have their own separate agendas for obtaining this common goal. For a female chimpanzee, the infant she bears will be one of only several she will produce in her life. Each infant must therefore receive as large and time-consuming a dose of motherly care as necessary. For a chimpanzee, this may entail four to five years of nursing, followed by years more of emotional and psychological dependency. A male, meanwhile, invests only a few minutes or hours of his life to mating with the mother of his babies; he provides no direct parental care whatsoever.

Although this picture might make you think chimpanzee society is utterly male-dominated, that is not necessarily the case. The schism between male and female mating tactics leads to some more general observations and predictions about chimpanzee society. Females' greater investment of time and energy predicts they should be choosier about their mates than their mates are about them. Female chimpanzees are active choosers, too. They sneak behind shrubbery to have sex with males other than the alpha in order to avoid being assaulted and beaten by the alpha. When a male chimpanzee wants to have sex with a female, he will give her a hand gesture as a signal of his intent and desire. A common signal is to reach out and grasp a nearby bush, shaking it rapidly. If the desired female ignores this gesture, he will shake it again, but more insistently. If repeated signaling does not get the desired attention from the female, the male may attack and brutalize her. The female nevertheless does not always submit to male mating intentions, the risk of physical violence notwithstanding. Female great apes of all four species are active players in their societies, not passive recipients of male mating tactics. One could say that females are the driving force in chimpanzee society, because so much of a male's energy seems to be directed at obtaining access to females.

And females mate quite liberally. Caroline Tutin, a chimpanzee researcher at Gombe in the 1970s, observed an individual female mating with up to half a dozen males in the same day when she was swollen. Why would a female be so promiscuous as to mate with most of the males in the community in such a short time span? She may be attempting to spread around the belief among males that each might be the father of any offspring that ensues eight months after the orgy. Females do need the protection of males at times—during territorial incursions by neighboring gangs of males, for instance. Females also must worry that the males of their own community may turn on their infants. Infanticide has been observed in some chimpanzee field studies (as it has in dolphins), although the circumstances and causes are less clear than in some other primates, in which infant-killing fits an evolutionary scenario. Mating with a variety of males may be a female's way of cementing bonds with males on whom they may need to rely in the future, or at least from whom they seek a neutral, rather than hostile, relationship.

Some studies have produced genetic evidence that females living in one community may try to have their babies fathered secretly by males from neighboring communities. In other words, females may live with one set of males, but also shop for genes elsewhere. This speaks further to the ability of females to be independent agents in chimpanzee society, not necessarily submitting to males. Rather than consider chimpanzee society to be male-dominated or female-driven, we should consider it as we consider our own society. There are spheres of influence, many of them traditionally controlled by men, but in many of which women exert control and influence as well.

Some species of dolphins don't live in stable groups but rather in fluid and flexible fission-fusion societies akin to those of their terrestrial ape counterparts. Bottlenose dolphins' social organization, for instance, is similar to chimpanzees in that adult males may form strong alliances with other males. Both female chimps and female bottlenose dolphins are seasonally polyestrous, have similar intervals between giving birth, and reach their sexual maturity at about the same age. In both species, males and females tend to aggregate when the females become receptive, and promiscuity seems to be the rule in the ocean as it is in the forest. Aggression and belligerence between rival males and strong male-male bonds are another communal aspect of both species, but in dolphins these bonds play a particular role, as I discover one morning during my field research.

I have just finished my breakfast about the time my assistant calls on the walkie-talkie from the fishing wharf. Her voice has an urgent tone. She tells me excitedly that two dolphins are harassing Bonita. Bonita is a bottlenose female frequenting the waters of Rio Lagartos. She has a distinctive black spot on her left "cheek" that reminds us of a beauty mark in humans. We have often observed Bonita foraging near the moored *pangas,* sometimes following the fishermen out to sea in the early morning.

When I arrive on the scene, my assistant gestures frantically, attempting to jam the summary of her observations into a few short sentences. I can see Bonita near a red *panga.* Behind her, a pair of dark dorsal fins follow closely. Her movements are panicky and abrupt as she maneuvers between

the moored boats, coming irregularly to the surface to catch a breath of air. She behaves evasively.
The males are herding her, synchronously chasing
her, charging her from both sides, turning upside
down at times and snapping at her pectoral fins.
They slam repeatedly against Bonita, cutting off
her every attempt at escape. The conflict continues
for a few more minutes until she finally manages
to evade the two and vanishes offshore to safety—
at least for today.

What Bonita just experienced off Rio Lagartos
is one of the various strategies bottlenose males
adopt to mate with females. In Sarasota, Florida,
and Shark Bay, Australia, associations between
males and females are tied to the female reproductive state. In Shark Bay, two or three males may
form alliances called consortships to mate with a
single female. These may last for only a few minutes or continue for months. A consortship may
begin with a simple chase that can evolve into aggressive behavior and displays that restrict the female's ability to escape or take up with another
consortship. To solicit sex and "convince" a female
not to leave them and copulate with another male

group, a consortship may produce low-frequency "pop" sounds. Males may also become increasingly more violent toward their intended sexual partner in an attempt to bully her into mating. As Bonita clearly showed us, the herding events are not always well received by females.

These types of consortships are called primary alliances and are the most simple. They consist of only a few males, generally occurring when the bonds among individuals are quite strong. These primary alliances function mainly to preserve the consortship at all costs, even with violence toward the female when necessary. Secondary alliances may occur when two primary alliances cooperate with each other to protect "their" females from external intrusions by other alliances. Sometimes, these secondary alliance groups may grow fairly large. As they grow, they tend to become less stable, because the bonds among individuals in these larger groups tend to be weaker. These larger associations are called super-alliances. In these, males may frequently switch companions to conquer females, and the group may become quite brutal with other nonaffiliated primary alliances if reproductive success is at stake.

Males in primary and secondary alliances are related to one other. The advantage of having some "family connection" in the same alliance is that a member will increase his own fitness (his genetic contribution to future generations) even when he is not able to mate. So long as one of his allies is, the genes will be maintained "within" the same alliance. In other words, if you can't have children but your brother or nephew can, your name, together with some of your genes, will be passed to the next generation. In a super-alliance, however, males may not be related at all. So what's the purpose of such a large group? Perhaps the real value of a super-alliance lies in its ability to easily subdue other alliances and steal their females, thereby increasing overall mating success.

How different, then, is male-male bonding in chimpanzees and dolphins? Some bottlenose dolphins form alliances to gain reproductive success, whereas chimpanzees use alliances to determine social rank and for success in aggressive intercommunity interactions. Aggression among adult dolphin males doesn't seem to have the same function of establishing dominance, as it does in chimps, although the role of dominance in wild bottlenose dolphin societies is not yet understood. Thus, the alliance mechanism likely goes with a large brain, but the differences in function have something to do with the diversity of the terrestrial versus the marine environment.

The strong male-male bonding and mate guarding common in bottlenose dolphin societies is not typically found in other dolphin species. Males in resident killer whale populations live in cohesive pods where they bond strongly with their mothers rather than with other males. They have never been observed joining forces in an effort to control females. Resident killer whales and bonobos are both known as polite neighbors, behaving peacefully within their respective communities. Even when killer whale clans overlap territories, they tend to mind their own business, showing no signs of aggression toward fellow communities.

Killer whale pods are truly matriarchal. A resident pod may be formed by one or more matrilines that usually travel together. Matrilines include a mature female and all her direct male or female descendents. Resident killer whales stay with their natal group for the rest of their lives. Their bonds are so intense that the animals remain together even after the matriarch has died.

They avoid inbreeding simply by mating with other clans, outside the matrilines. Resident killer whales, in fact, always choose partners with a different "language" or "dialect." In other words, they seek out animals that don't sound at all like themselves. In the killer whale world, the more similar the dialect, the more closely two individuals are related.

Transient killer whales are another story. They are entirely distinct from resident whales and there is no exchange of members or interbreeding between the two. The relationship between mothers and offspring is also different in transient killer whales. From birth, both male and female offspring stay with their mothers for some years. This maternal bond, however, may dissolve at the time of their sexual maturity. At that stage, males will either remain with their mothers or leave them to become solitary individuals that form labile associations with other groups. Females usually leave their mothers at sexual maturity to aggregate with other groups. In their late twenties they will either form their own pod, or return to their mothers.

The striking difference in association patterns between resident and transient killer whales may be explained by their very different diets. Transients feed mostly on marine mammals, for which the optimal size of a hunting group is about three whales. As the group size increases, the prey success rate decreases, so a small group size rather than a larger matriline is preferred. In resident fish-eating whales, competition for food is no big deal, so the offspring have no good reason to abandon their natal group.

What about female-female bottlenose dolphin relationships? Female social life seems to vary broadly within and between populations; even more so than in female chimpanzees. Some dolphin females are solitary, while others spend most of their lives in coalitions with other females called bands, which serve to protect against aggressive males. These bands are particularly well studied in Sarasota Bay, where researchers have noted re-markable stability in these female associations. Even over the course of many years, they have observed very few changes in composition of the initial band structure. But why is there such variation in female social grouping? Female reproduction is limited by their access to resources, so their social grouping patterns are greatly influenced by ecological factors. The patchy distribution of prey in the ocean and the risk of predation are the two most important factors that determine how female-female relationships develop. Sponge-carrying females in Shark Bay, for instance, spend a lot of time alone looking for prey and don't seem to need a band for their daily survival. On the contrary, Sarasota females favor the company of others, which may offer protection against calf predators or male aggressors. In a female band, a dolphin like Bonita might have been safe and sheltered from the troublesome males sneaking around the *pangas* of Rio Lagartos.

When we look at the strong relationships between a mother and her offspring, we find that chimpanzees and bonobos are quite similar to their aquatic cousins. The birth of a bottlenose dolphin somewhere in the Pacific is accompanied by the same level of care, excitement, and concern from the mother and her female associates as the birth of a newborn chimp in the middle of an African forest.

Young bottlenose females, for instance, have an intense interest in the birth process and what follows. It is their chance to learn firsthand how to care for an infant. When the time comes, they have already prepared for motherhood. Babysitting is widespread in both primates and several dolphin species, but it is taken a step further by bottlenose dolphins.

Infant chimps at the Gombe Stream Reserve in Tanzania never wander more than a few meters from their mothers during their first year of life. Dolphin newborns, however, gain some independence after just two months of age. The bond between bottlenose mother and calf remains primary, but other females, called "aunts" by researchers, are allowed to take over momentary responsibility for the calf, sometimes escorting it for a swim of up to a hundred meters away from its mother. Child-rearing is as expensive in terms of time and effort for primates (including humans) as it is for dolphins; so, alloparental care by other members of a group may be extremely useful. In our modern human world, alloparenting is becoming less prevalent, but in the hunter-gatherer societies of our past, along with some traditional societies of today, sharing childcare among the members in a group was and is still considered the norm.

Despite the extra parenting help, however, sometimes only weeks after birth a precocious dolphin calf may become so self-reliant and fearless as to venture off by itself without a baby-sitter. One summer day, from a *panga* along the coast of the Gulf of Mexico, I was tracking Torta and Ben, a mother and infant pair of bottlenose dolphins. At the time, Ben was less than one month old and hadn't yet grown comfortable with the process of breathing. Instead of the fluid regularity of his mother's breathing, Ben would surface at irregular intervals, jutting half of his body out of the water to catch a breath, though always remaining at his mother's side. We moved closer to the pair and stopped the boat to observe. Fascinated by our *panga* and the strange individuals in it, Ben left the security of Torta to explore. At that moment Ben's actions reminded me of a typical human tot's in-

quisitiveness about the world. He possessed a desire to discover things for the first time, and the
awkward movements of one still unaccustomed to
his new life on Earth. At first, he examined our
outboard engine, which was making weird noises
and producing a huge number of bubbles. He
seemed fascinated by the spherical shapes full of
air frothing the surface. Then, turning toward the
bow, Ben swam slowly along the port side, almost
touching my arm with his rostrum as I leaned over
for a better look. All this time, Torta remained
still, maintaining her distance. She was vigilant,
constantly turning her head to check on Ben, but
made no move to swim over to us. She was allowing her infant to take its first steps to discover
this new, bizarre world of outlandish creatures by
himself.

The water is a bottlenose calf's first playground.
It can gambol with its playmates and snack on fish
at will, always having mom nearby. Calves have a
lot to learn from their mothers before taking their
own road, but even within populations, there seems
to be great variability in the time they spend with
and away from their parent and with or without

alloparental care. Killer whale mothers, on the other hand, are not aided by other females—they parent alone. Taking their calves' education a step further, they invest great effort in tutoring their offspring with a high degree of inventiveness in their teaching techniques, as one research crew discovered.

It was late evening on January 15, 2006 aboard the *National Geographic Endeavour*. The ship was passing just above the Antarctic Circle when a small pod of killer whales caught the attention of the research crew. What they observed had never been seen. It was a highly organized hunting technique, carried out with the deadly precision of a professional assassin. A pod of grey and white killer whales was moving slowly toward a seal, which lay on a slab of floating ice. In a highly organized effort, the whales swam at the ice at high speed, veering off at the last moment to create a wave big enough to jar the ice floe and wash the seal off its safe station. A few young whales remained off to one side, attentively monitoring the entire operation. Their role, it seemed, was to observe and learn the strategy used by the adults and to remember it for their own use in the future. It will be up to these young whales, when they reach the adulthood, to pass along the same successful hunting technique to the next generation—one more skill that may prove vital to the group for their survival in the freezing waters of Antarctica.

When a dolphin calf reaches a certain age or when the mother becomes pregnant again, she may decide to release her infant once and for all, no matter how attached the young dolphin might be. If the calf is unwilling to leave, this maternal separation may take on an aggressive facet, and the calf may end up on its own, carrying nothing from its mother other than a few unsympathetic tooth marks. But separation is a necessary step. After it has learned the required social expertise and foraging techniques, the young dolphin must now become skilled at how to catch food and survive on its own. Usually the young dolphin is in good company; other offspring are also taking independent steps out into the ocean, where they will begin to experience the next phase of their lives.

A bottlenose dolphin's social life can vary substantially from individual to individual depending on whether a highly social mother or a rather solitary one raised the calf. In a highly social environment such as the waters off Sarasota, for example, juveniles spend lots of time playing with each other while developing their social skills. Play seems to be very important in a young dolphin's life, but its meaning is still poorly understood. Play may strengthen social skills for use in later life or reinforce the bonds among individuals. Males are the most involved in these sociosexual games, exchanging roles like Hollywood actors and attempting to mount females (and sometimes each other). This is also when juvenile males begin to establish a sort of hierarchical dominance, a juvenile expression of power and control over their peers that seems to be of less importance in adult life.

There are numerous parallels between dolphins and apes in sexual alliances, mother-son bonds, hunting, and territorial aggression. Are these coincidences or the result of similar factors coming together in two distantly related but equally big-

006

brained, intelligent animals? What common factors in environment, social behavior, and brain size produce such a sophisticated degree of intellect in two groups of animals that are otherwise different in almost every way? To understand this we must turn to the cognition of both creatures, and consider the evolution of intelligence.

chapter five

COGNITION: MINDS IN THE
SEA AND FOREST

INTELLIGENCE IS PERHAPS the rarest commodity in all living things. It is very hard to define, and even harder to find in the animal world. Of all the 5 billion or more species, only a handful have possessed a high degree of intellect; apes and humans (including many extinct forms of both), dolphins and whales, and perhaps elephants. That is the brain-power short list, and it is short indeed. So we must begin by asking, "What's so beneficial about having a big brain?" The answer to this question leads us toward understanding why these lineages in particular, apes and dolphins, have such oversize brains.

Scientists have argued for years about how to define intelligence. As debated as the value of IQ scores are for people, the usual intelligence tests cannot be applied to apes or dolphins at all. The term "intelligent" has very little precise meaning when we speak of animal behavior in particular; does it refer to learning, memory, or some other cognitive factor? Researchers in the area of primate and dolphin intelligence consider problem-solving capacity to be a measure of intelligence. Problem-solving cognition is a successful adaptation in these animals because it allows them to respond effectively to novel situations.

Every day, both our ancestors and modern non-human primates must navigate their way through an environment, both physical and social, that tests their ability to survive and reproduce. Primate intelligence is the way natural selection chose to promote those skills. You might think that some animals stumble on their food, but most have keenly honed strategies for food-finding. Bumble bees and hummingbirds travel across a field of wild-flowers in an up-and-down series of rows, travers-ing an efficient search grid. Chimpanzees travel from tree to tree feeding all day long; do they re-call the locations of thousands of food trees, or do they travel a path that takes them randomly into food sources? Many fruit species, such as figs, ripen unpredictably, but a party of chimpanzees will usually be at the tree as soon as ripe fruits ap-pear. This suggests that the apes are monitoring the fruiting status of trees as they forage, and re-member which trees are worth waiting for.

Dolphins must cope with a similarly complex environment. Although it might seem that the ocean is nothing at all like a rain forest, from the perspective of a food-finder they may not be all that different. The ocean is a vast place, with widely scattered resources that are constantly on the move. Dolphins seek schools of fish, and apes seek stands of trees laden with ripe fruits. Unlike the fish, the trees stay in one place, but the temporary, unpredictable nature of fruit availability makes the apes' favorite food an equally hard-to-get resource in a complicated forest world.

Evolution may have placed a premium on flexibility and the ability to remember and predict based on past experience the best spots to find schooling fish and the time of year when the best fruit trees are ripe. This ecological theory of intelligence extends this logic to early humans, who also had to cope with a complex environment.

Even though the physical environments dolphins and chimpanzees live in are complex, their common social environment is far more intricate. It is in their dealings with other members of their group that both make the best use of their exalted brain power. The prevailing view among scientists today is that the brain-size increase that occurred in great apes and was extended into early humans resulted from the premium natural selection placed on individuals that were socially clever. This theory of social intelligence argues that the primary evolutionary benefit of large brain size was that it allowed apes and hominids to cope with and even exploit increasingly complex social relations. In large social groups, each individual must remember the network of alliances, rivalries, debts, and credits that exist among group members. This is not so different from the politics of our own day-to-day lives. The primatologist Frans de Waal has observed that chimpanzees seem to engage in a "service economy" in which they barter alliances and other forms of support with one another. Richard Connor, a marine mammal expert, discovered that male bottlenose dolphins do something similar, bonding in tight, cooperative alliances.

Those individuals best able to exploit this web of social relationships would likely have reaped more mating success than its group mates. The ability to subtly manipulate others is a fundamental aspect of group life. Some researchers even argue that as average group size increased, the neocortex of the brain increased in size to handle the additional input of social information, in much the same way that a switchboard would be enlarged to handle added telephone traffic. This effect holds true even when the evolved patterns of social grouping are taken into account.

The Pinnacle of Intelligence

Both apes and dolphins possess extraordinarily large brains compared to the size of their bodies. In the case of both, this is because the neocortex—the cerebral package that sits inside the globe of your own skull—is so elaborately developed compared to that of lower animals. Dolphins, in fact, possess a larger neocortex than that of any primate, including humans, and the ratio of their brain volume to their body size is larger than that of any primate, except humans. Merely in terms of brain volume compared to body size, therefore, great apes and cetaceans are at the apex of brain power in the nonhuman animal world.

Using brain volume as an indicator of intelligence is problematic, however, since dolphins and people use their brains for different tasks. The overall size and weight of a dolphin's brain is somewhat similar to your own. But much of a dolphin's brain is devoted to making and understanding sounds, both vocal and sonar. The organs and space devoted to the sonar apparatus account for a lot of the size difference between dolphins and other mammals. Dolphins produce sounds in their nasal cavity, but process and amplify them in a fatty mass of tissue called the "melon" next to the nasal cavity in their forehead. The organization of a dolphin's brain is superficially similar to an ape's; the frontal lobes resemble those of people, and the temporal lobes house a structure that resembles the language center of the human brain. Dolphins and primates have not shared a common ancestor for more than 90 million years, yet they possess cognitive abilities that are strikingly convergent. Still, many researchers feel these similarities are superficial, and caution that we should not leap to consider intelligence a prerequisite for the social complexity we see in apes and dolphins.

What are the fundamental attributes that would place dolphins and great apes near humans, at the pinnacle of what we call the "intelligent" world? The abilities to be flexible, remember, imitate, use tools, understand language, and be self-aware are certainly among the things that make humans extraordinary. But might some dolphins and great apes possess such attributes?

Technological Smarts

Until a few years ago no one would have guessed that an ordinarily underappreciated invertebrate like a sponge would have achieved much notoriety, especially among scientists expert in the study of cetaceans.

It started in 1984 at Australia's Shark Bay where Halfluke, a female Indian Ocean bottlenose dolphin, was observed carrying a large, cone-shaped sponge on the tip of her rostrum like a mask. Following this first and seemingly unique sighting, researchers began to notice that not only Halfluke but other dolphins exhibited this kind of behavior as well, always in the same deep-water channel of the bay. The dolphins were carefully inserting their noses into the cone-shaped sponges. Water pressure created by the dolphin's fluid movement held them in place. Not all the bottlenose dolphins in the group wore these soft nose-caps; only a few solitary individuals were observed over a period of several years. Researchers called these mostly female dolphins the "regular sponge carriers."

From the perspective of a human researcher, it was rather challenging to decipher how a sponge that covered a large part of the dolphin's face and probably interfered with its ability to open its mouth could have been of any advantage to the animal. Were the dolphins using the sponges to play? There were other instances, elsewhere in the world, in which wild bottlenose dolphins had been observed carrying objects like a kelp string or a piece of plastic; and it wasn't news that dolphins sometimes amused themselves with simple toys. Carrying of sponges, however, was a stereotyped and meticulous behavioral pattern displayed by only a few solitary females who practiced it for hours on end and repeated the behavior over a long period of time. This was certainly too methodical to be just play; it had to be something else.

So were the dolphins perhaps extracting some special antibacterial or antifungal substance from the sponges? Several species, including our own, are known to use plants and other objects extracted from the environment as medicine. The idea of a therapeutic sponge used by dolphins sounded good but left too many questions unanswered. Why, for instance, were only a few seemingly healthy females interested in this marvelous medicine? And why did they use it regularly over a several-year span?

The lack of answers led researchers to seek a third and more attractive hypothesis. A wide array of bottlenose dolphin foraging strategies had already been documented in Shark Bay, but it was becoming apparent to researchers that this group of sponge-wearing females were doing something altogether different. The sponge-carrying females were usually spotted while foraging or echolocating in search of prey. While toting the sponges, they were pursuing prey on the sea floor and casting off their nose caps at the last instant before the capture of their food. Might the dolphins have been wearing the sponges as a shield? It now seemed that the sponge mask was an intelligent way for the females to protect themselves from a variety of harmful and toxic organisms near the sea floor and to avoid the abrasive sand, rocks, and broken shells that littered the deep waters of the channel. By using the sponges, the dolphins may also have been able to agitate the sea bottom and flush out burrowing prey species from the sand. This behavior seemed to be a "tradition" vertically transmitted from mother sponge carriers to their daughters.

We use tools like spoons and chopsticks in our everyday life, but could sponges be considered a kind of foraging tool for dolphins? To answer this question, we need a rigorous description of "tool use." The most accepted definition was given by biologist Benjamin Beck in the 1980s: "tool use is the external employment of an unattached environmental object to alter more efficiently the form, position and condition of another object, another organism, or the user itself when the user holds or carries the tool during or just prior to use and is responsible for the proper and effective orientation of the tool." Using this definition, the dolphin's use of sponges for protection against stinging organisms and sand abrasion or for extraction of prey from the sea floor would meet all the criteria for using a tool, making the wild Indian Ocean bottlenose dolphin the first tool-using cetacean ever documented.

We're sitting around a massive earthen termite mound in a forest in Tanzania in the same way a bunch of conventioneers would crowd a buffet table. Three of our group of twelve are human observers; the other nine are chimpanzees. A termite mound houses a feast, especially during the start of the rainy season in November. The red clay soil of the mound is moist, and the myriad tunnels the solider termites have made can be easily picked open by chimpanzee fingernails. A female and her two daughters sit on the slope of the mound, the mom looking intently for tunnel openings while her daughters just as intently watch her. A few feet away an elderly male is picking at the mound. He breaks off a bit of soil and brings his termite-extracting tool into play. A twig he had snapped from a bush a few minutes earlier is tucked sideways in his lips. He plucks this from his mouth, clips off the frayed end in his teeth, and pokes it down the tiny tunnel. A few gingerly pokes later, the probe comes out of the mound coated with brown termites like a wooden spoon dipped in chocolate. The chimp runs the twig through his lips and munches the insects into a protein-rich pulp. Lunch.

Watching him expertly fish for the termites, I get an idea. Picking up one of his discarded twig tools, I sit down on the opposite side of the mound, within his line of vision, and begin to do my own fishing. It's easier than I thought; on the first try I extract a stick full of angry, oversized soldiers, their jaws snapping at me. The chimpanzee immediately sees this, stops what he is doing, and moves to my side of the mound. He walks over to "my" termiting tunnel (I was proud to have located such a productive vein on the first try), looks at me, and then down at the mound. As I begin to back away—we try to maintain a several-meter distance between apes and humans at all times because of the apes' vulnerability to nearly every human disease—he comes closer, almost pushing me off my spot on the earthen mound. With a wave of his hand, I know I am being dismissed; he takes up a tool and begins to fish from my tunnel.

Jill Pruetz and colleagues recorded another example of a chimpanzee finding the right tool for the job. They recently observed a chimpanzee using a stick that it had peeled at one end to a tapered point in order to jab it into a tree cavity. Inside the hole was a bushbaby, a squirrel-sized animal that the chimp then extracted, apparently dead or injured, from the hole. The chimpanzee had foreseen a problem in immobilizing and extracting its intended prey and thought up a clever tool solution—albeit not exactly a spear—in response.

This sort of technological intelligence is one of the behaviors that made chimpanzees famous in the early 1960s, when Jane Goodall reported apes manufacturing tools of grass blades and twigs with which they fished termites from their earthen tunnels. Ironically, Darwin had proposed technological competence as the key to the rise of intelligence in his 1871 book, *The Descent of Man*. He saw a feedback loop between the emergence of upright posture, tool use, and the evolutionary expansion of the brain. Once ape ancestors began to walk upright, Darwin postulated, their hands were freed to learn to make and use tools. The best tool makers would have reaped a survival and reproductive advantage, which would have placed intelligence at a premium. Thus our ancestry would have hinged on the ability to be technological, with its accompanying posture and cerebral expansion.

Darwin's hypothesis, while logical, has since been revealed as flawed. Modern geology has dated the origins of tool use at least 3 million years after the transition from quadruped ape ancestor to bipedal hominid. Our earliest hominid ancestors, the australopithecines and their immediate predecessors, arose around 6 million years ago based on current evidence. But the first evidence of stone tools does not appear until 2.5 million years ago. Moreover, although brain size increased slowly and incrementally throughout our lineage's history, the size of the cerebrum did not really explode until only the past few hundred thousand years; before that time brain size increases simply scaled to body size increases. So the three key events that Darwin saw as interconnected—upright posture, tool use, and brain size—were in fact disconnected by millions of years.

The Sophisticated Art of Imitation

A key piece of evidence that apes and dolphins are capable of humanlike thought is their ability to imitate others. By imitation, we mean the ability of one individual to copy the behavior of another. You might think that imitation would be simple to recognize. But as psychologist Richard Byrne points out, scientists have tended to define imitation by what it is not rather than by what it is. Only after excluding lower-order forms of copying do we call a behavior imitation. Many birds, for instance, are accomplished mimics and have extensive, highly accurate repertoires of songs of other species. Mockingbirds not only sing songs of other bird species, they also mimic the sounds of cats meowing, rusty gates creaking, and kettles hissing. Parrots of course are masterful mimics of human speech. But there is very little evidence that parrots understand the content they mimic (with the possible exception of Alex, an African gray parrot who was the subject of research for many years).

Determining whether apes understand what they appear to be imitating is at the heart of the debate over the role of imitation in ape intelligence. Skeptics point to two key criteria for declaring that apes and humans think about problems similarly. First, apes must have a theory of mind. By theory of mind, I mean the ability to place themselves in the mind of others, to intuit others' mental state, and to act on that insight. Second, apes must be able to copy an entire sequence of behaviors, each step of which may serve little function, putting them together into a whole behavior sequence. This requires not simply the ability to repeat a sequence of behaviors but also the ability to recognize each step. Many animals can imitate simple actions they observe performed by humans or by members of their own species. Captive chimpanzees have readily copied tool use behavior they observed performed by other chimpanzees, even though the same tool use performed by a human did not stimulate copying.

To step into the mental shoes of another human is one thing; to recognize it in a nonverbal animal is quite another. How do we know that when an ape copies a physical action, it truly understands what it is copying? Researchers usually make this judgment by distinguishing apparent imitation from trial and error. Other researchers have attempted to distinguish between imitation, which they regard as a uniquely human ability, and emulation, which they regard as an ape's approximation of human imitative abilities. To these researchers, imitation is copying with an understanding of both the goal and each step toward it. When an ape sees a person place a bandage on a cut, and then copies this behavior by placing a bandage on its own arm, is this imitation? Some scientists regard this as emulation, not imitation, because there is no evidence the ape understood the meaning of the act of placing the bandage. Whereas a human child who watches a tool being used gains an understanding of deeper causation, the chimpanzee lacks the ability to understand the causational aspects of the process.

The psychologist Michael Tomasello set out to determine if a chimpanzee could imitate the way a person performs a task requiring forethought and planning. He set up an experiment in which a ball could be reached at the end of a table only by inserting a rake through a grate from the opposite end, so that the ball could be dragged toward the person holding the rake. When a small child observed this, it took the child only one trial to learn how to get the ball using the rake, and the child perfectly imitated the researcher's demonstration. The chimpanzee was also able to get the ball after watching only a single trial. However, the chimp devised its own method of using the rake to obtain the ball, not the style the researcher had demonstrated. Tomasello concluded from this that the chimpanzee failed to imitate the process—even though it could achieve the same result—because it lacked a theory of mind that is necessary for true imitation. Tomasello labeled what the chimpanzee did as emulation; achieving the goal without understanding the importance of imitating the process.

Tomasello's experiment and others like it have been carried out by researchers who doubt that chimpanzees possess a theory of mind. But critics point out that the rake-ball test is too similar to games children play every day, while utterly unfamiliar and unnatural to chimpanzees. This is a persistent problem for laboratory studies of great apes. Their cognition evolved in response to the pressures of tropical forest environments, not laboratories. Even the best laboratories provide impoverished social learning environments for their study subjects compared to what children experience growing up. No human orphan who had been traumatically separated from his or her mother as a toddler and then reared in a socially impoverished environment would be deemed an appropriate model for studying normal childhood cognitive development. Apes, psychologically and emotionally similar to young children of similar age, may be no more suitable as models.

Bottlenose mother LulaMom and Nin, her newborn calf, are swimming side by side less than a hundred yards from the beach at Malibu. Nin, barely a month old, is dark in color with a flabby dorsal fin. Nin still bears the fetal folds from the womb, which appear as light stripes on the sides of his body. Swimming in his mother's "slipstream," the hydrodynamic wake created by her movement through the water, Nin catches a free ride through the waves as they break near a shoreline packed with neoprene-clad humans on surfboards. The calf observes intently as LulaMom catches a few fish near the surface. Then, it's Nin's turn; imitating LulaMom's movements, Nin clumsily disappears under a wave, presumably chasing a fish. A moment later, unsuccessful, he bursts out of the water to breathe with his whole head exposed, having not yet learned the fluid and effortless porpoising motion of adults. Nin tries again, following mother's example, until success is finally attained, then resumes his favored spot in her slipstream. By sticking close to his mother, constantly observing and repeating his mother's behavior, he will become independent and able to survive on his own. Does Nin simply mimic the actions of his mother, eventually learning by doing that hunting leads to eating, or is there more to it?

When we speak of imitation and theory of mind in dolphins, we enter an area filled with controversy and confusion in the scientific community. Imitation, for instance, may be viewed at different levels, from simple or action-level imitation, in which movements or actions are copied without understanding, to true imitation, sometimes referred to as "emulation" or "program-level imitation," which implies that the imitator has an understanding of the actions he or she is imitating. A recent study conducted by Wendy Fellner and her colleagues points out how the synchronous swimming of a calf near its mother during the first months of its existence has advantages over and above reducing the hydrodynamic costs and the potential risk of predation. This close contact may also play a key role in initiating the process of simultaneous imitation of other dolphin behaviors. Following this thought, Nin will pass from the stage of action-level imitation during his years of development to a more complex program-level imitation in his adult life that will help Nin exist within a multifaceted societal group.

For dolphins as with apes, a similar set of

problems plague the study of cognition. Most of the experiments attempting to prove or disprove a theory of mind in these animals come from close, unnatural, impoverished environments, where dolphins are not required to make full use of their brain power and where the socioecological soundness of the experiments may be in doubt. That said, there have been many observations made both in captivity and in the field that may indicate something more complex than simple mimicry or action-level imitation in dolphins.

To assess a dolphin's ability to imitate, there are some significant physical limitations to consider. In an experiment with a chimpanzee, we might ask him to imitate a human gesture, like raising a leg. It is easy to see if he imitates us or not, given the similarities of our body features. Now imagine the same experiment carried out with a dolphin. When we raise our leg, what does a dolphin do? This is no simple request for an animal that doesn't have arms or legs. To imitate our action, the dolphin needs to translate the act of raising a human leg into something it can express with its own body and then perform the task with that attribute. In experiments like this one, dolphins have been observed raising a flipper or their entire tail out of the water in response to the human request. This act requires much more than the simple copying of a physical action; it implies some level of representational sophistication.

The research team of C. K. Taylor and G. S. Saayman documented some interesting examples of dolphin behavior in captivity that seem to take us another step closer to the existence of a theory of mind in these animals. In one anecdote, a trainer was smoking a cigarette near the underwater porthole of a dolphin tank. A calf stopped in front of the porthole and watched the trainer blowing smoke in the air near the glass. After a while, the calf swam away to its mother waiting nearby, nursed, and then, with a mouthful of milk, returned to the porthole where it first saw the instructor. The calf slowly released the milk from its mouth, producing a white cloud similar to the cloud of smoke produced a few moments beforehand by its biped counterpart.

Another account involved a dolphin and a scuba diver sent into the dolphin tank to remove the algal growth from the underwater portholes. Following next to the diver, the dolphin observed the cleaning activity with great interest. After monitoring the diver's behavior for quite some time, the dolphin picked up a seagull feather in its mouth and started "cleaning" the windows with it while releasing sounds and streams of bubbles similar to those emanating from the diver's scuba gear. Perhaps not entirely convinced it had finished the job, the dolphin then collected stones, paper, and dead fish and used them in the same way. This behavior seems more than just an action-level emulation; the dolphin was able to understand the relationship between the actions of the diver and the outcome. Many other examples of imitative behaviors such as these have been recorded in the literature—so many that we should consider looking at them as having some scientific significance.

The existing evidence suggests that some dolphins are able to imitate and remember postures and movements originating not only from behaviors already familiar to them but also from novel activities displayed by other animals, including our own species. In general, adult dolphins seem

to learn new actions more easily than young individuals, but young dolphins, like human children,
seem to learn quickly when something catches their
interest.

There are also a few cases of vocal imitation
in dolphins recorded both in captivity and in the
wild. Bottlenose dolphins, for instance, are fantas-
tic imitators of manmade sounds, as observed by
Peter Tyack of Woods Hole Oceanographic Institution. They can listen to a sound and repeat it
with exceptional accuracy within just a few seconds of hearing it. In the wild, dolphins are able to
address each other from a distance when they get
separated, a behavior called "whistle matching," first
observed by another Woods Hole researcher, Vincent Janik. An individual might echo a companion's call as a welcome greeting or as an aggressive
warning, depending on the situation.

Other examples of vocal learning come from
killer whale studies. David Bain described a captive
killer whale from Iceland that was able to repeat
and then learn the vocal repertoire of a British
Columbian tankmate over a three-year period. A
few years later, another researcher, John Ford, re-

ported what he calls "interpod call mimicry" in the wild, showing again that orcas are capable of vocal learning. Vocal imitation seems to be extremely important in the communicative world of a wild dolphin.

Dolphins can imitate postures and sounds with differing degrees of complexity. Observations indicate that dolphins may understand more about their actions than can be explained by simple mimicry of behavior. Is this then indicative of the existence of a theory of mind? Certainly more studies are necessary to elucidate these questions, but with dolphins as with the study of apes, we must accept that a theory of mind in these animals may not closely resemble a theory of mind in humans.

Look Who's Talking

We are offshore in Santa Monica Bay, California, following about 200 short-beaked common dolphins. The otherwise calm surface of the ocean is teeming with what seem to be white caps formed by the porpoising motion of their fins at the surface. They are hunting for food, deployed like a platoon of soldiers on a battle line. They move steadily northward, spreading out to cover a larger body of water, keeping us ever busy in recording their behavior. A few whistles reach the headphones of my assistant, who is leaning over the DAT recorder intently listening to the signal coming through our hydrophones. Suddenly, beneath the surface, something happens. My assistant waves me over for a listen and motions to the captain to move the boat closer to the group. I put on the headset and listen as the "platoon" divides into smaller units who spin off in different directions. The occasional whistles are gone, replaced by an intense cacophony of whistling and rapid click trains.

Half an hour later we are still recording as the school of dolphins feeds energetically on a feast of sardines seven miles off Los Angeles. Now, all we see are dorsal fins and flukes at the surface and all we hear are streams of high-pitched clicks. I am photographing their behavior when suddenly the clicking stops, turning instead to noisy and intense whistling. When I look back into the viewfinder, the surface is calm and the dolphins are gone. My assistant tells me that the whistling has stopped as well. I scan the water through the camera lens for a sign of them and stop on a large, triangular fin with a light-grey patch on the lower front side. I look up just in time to see the great white shark as it passes under our boat.

I don't know what the dolphins were "talking" about just before we spotted the shark but it was probably not just about how great the meal was at this particular feeding spot. Their language, or whatever we decide to call it, probably enabled them to advise each other of an imminent danger and keep track of where the other group members were at that moment.

What we know of the natural world comes from our five senses. What dolphins know, or might know, about their own environment comes mostly from their visual and acoustic sensory channels, which are highly developed and specialized for the aquatic medium. It is through these systems, particularly their hearing ability, that a dolphin can perceive its surrounding world and interact with its companions in a coordinated and complex social organization.

Dolphin sounds can be divided into three main categories: whistles, burst-pulse sounds, which are used for social communication, and clicks, which are used to orient and navigate in the water and search for prey. Dolphin whistles represent some type of specific code involved in the exchange of information between a sender and a receiver. They are somewhat similar to human whistles, except that the pitch, or frequency, is much higher.

These sounds are believed by many scientists to be a form of language. In the late 1950s and 1960s, John Lilly, a neurophysiologist, set out to break the interspecies language barrier. He believed that dolphins not only possess sophisticated linguistic abilities, but that they might also have the capacity to communicate with humans. His hypothesis was based on the fact that bottlenose dolphins have an exceptionally large brain with a highly developed neocortex, giving them the capacity, among other things, for the development of a natural language. After dissecting many dolphin brains, which made him a target for the antivivisectionists of the time, and spending days and days listening to dolphins speaking what he called "delphinese," his intrepid quest to break the language barrier and teach English phonemes to dolphins failed. As I mentioned in Chapter 1, however, he thoroughly captured the imaginations of the general public in the process.

Lilly wrote that "scientists should learn how to communicate with dolphins to prepare for communication with intelligent life in outer space," which unleashed rivers of criticism from the scientific world and a surprising public infatuation with his books. Those books painted an image of animals more intelligent than a man could ever be. Though all his hypotheses were negated, an important observation did emerge from Lilly's work, almost accidentally. While listening to a set of tapes of dolphin vocalizations, he realized that one of his subject animals was mimicking what the human researcher was saying during the experiment in a "high-pitched Donald Duck, quacking-like way." Lilly's experiments set the groundwork for a milestone series of studies on dolphin vocal learning, mimicry, communication, language comprehension, and cognition.

Following in Lilly's footsteps, Louis Herman and his colleagues at the Kewalo Basin Marine Mammal Lab in Honolulu began devising two artificial languages they hoped to teach to the bottlenose dolphins housed at their facility. The languages, one computer-generated with high-pitched words, the other, a sign language conveyed by arm and hand signals from a trainer, didn't approximate human conversation but, like our language, were based on a set of grammatical rules.

In an underwater classroom, dolphins Phoenix and Ake were taught a series of sentences. A trainer would say or sign "Phoenix Ake under," which meant that Phoenix needed to understand the command and then accomplish the task of swimming under Ake. The "students" showed remarkable academic promise and soon became able to distinguish between "right hoop left Frisbee fetch" (meaning "take the Frisbee on your left to the hoop on your right") or the opposite, "left hoop right Frisbee fetch." Herman concluded that his dolphins were able to understand both the meaning of the words and the sentence patterns, basically comprehending the semantic and syntactic components of the language-like instructions.

The dolphins' responses to the training were so successful that they were even able to formulate their own logical and spontaneous responses to meaningless phrases. For instance, when a trainer would occasionally "say" something that didn't make any sense, Ake ignored the command. Herman interpreted this to mean that the dolphin was not only capable of understanding words and sequences of words, but also clever enough to make decisions about them.

The experiments became more and more challenging for the dolphins. Trainers might appear on an underwater TV screen or with bodies obscured, leaving only the arms out for the dolphins to see. The goal was simple: to determine whether the dolphins could interpret gestural communication that included less and less detail. Even with the minimal, conceptual presence of communication, the dolphins were almost always able to give the correct response, leading Herman to conclude that these dolphins could make sense of artificial grammar.

Semantics and syntax have always been considered the key attributes of any human language. Our language can be considered syntactic in the sense that messages consist of components, each with their own meaning. Animal communication has always been considered to be nonsyntactic because their signals refer to "whole situations." In the evolution of our own language, the transition from the nonsyntactic to the syntactic form of communication was a defining step that would separate our human expressive ability from that of other animals. But by being able to acquire at least some of the proprieties of our syntax, Herman's dolphins taught us that the gulf between our respective forms of communication is not as wide as previously thought.

In the dolphins' underwater world there is an open system of communicative signals. For instance, there are the individual-specific and distinctive signals of the bottlenose dolphins, which live in fission-fusion societies, and there are the group-specific vocal repertoires of killer whales, which spend their existence in stable pods. Social signals and social structure in a dolphin group are interrelated in such complex and flexible ways that we still labor to understand what is really going on inside the big brains of these magnificent and intuitive animals.

The Ape Speaks

After so many years spent watching wild apes, the opportunity to have a conversation with one was a shock. I shouldn't have been shocked; the chimpanzees at the Chimpanzee/Human Communication Institute on the campus of Central Washington University are world-renowned. Washoe, Loulis, Dar, and Tatu have lived together as a social group for decades. They were first taught sign language by Allen and Beatrice Gardner in the 1960s. Washoe was the pioneering research subject, and Roger Fouts, the Gardners' graduate student, began a lifelong project teaching Washoe and her compatriots and then learning from them.

Today the four chimpanzees live together in a comfortable indoor-outdoor zoolike enclosure, tended to by a legion of nurturing students, care-givers, and scientists. When I visited them one day in 2003, they were a picture of contentment. Washoe sat idly reading *Newsweek,* casually flipping pages with her finger the way a girl at the beach reads *Cosmopolitan.* Occasionally she used her tongue to turn the pages. Now in her late thirties, Washoe has assumed the status of matriarch among the four chimpanzees. Loulis is her adopted son, brought to her as a tiny infant to replace her baby, who had died. Over the years, the four have astounded Roger Fouts and his research partner and wife Debbie with their linguistic sophistication. In the early days, Washoe acquired her signing skills the way any mute child does; by listening to and watching a parent. The Fouts were Washoe's devoted foster parents, and they reared her in a human home as they had raised their own children. Although there are critics who claim the signing chimpanzees are only parroting their trainers rather than understanding the concepts they sign about, the Fouts have provided compelling counterevidence. The four chimpanzees don't simply sign "apple" when shown a photo of an apple by a human trainer. They sit in their social group, observed only by the omnipresent videocameras, and sign to each other. Often it's just a trivial, two-word sentence, such as, "Hey, give me that apple." But it's conversation, probably not unlike what early humans used, and it happens without training or prompting or rewards. Such simple conversation is a wonderful example of both language and culture in action; learned traditions are passed down from one individual to another, encouraging an even greater range of communication. The limits of their language, comparable to that of human toddlers, don't change its overall importance. After all, in any daycare center there are a thousand interactions a day of lower communicative content and caliber. When we see simple language in children, we still label it language. We know it will someday develop its full range of complexity. When we see the same process in chimpanzees, we shy away from drawing the same conclusions.

The Central Washington University signing chimpanzees have taught us a great deal about language itself. For instance, Washoe and her mates use one sign for "dog," which they apply to Dachshunds, Collies, and Great Danes alike. In other words, "dog" is a natural category in the chimpanzee mind, and the sign they use is chosen accordingly. The Gardners and Fouts showed years ago that signing chimpanzees use novel signs to describe objects for which they know no other sign: hence the signs for candy and for fruit, given together, to refer to a watermelon. Although most chimpanzee sentences are only two signs long—much like a toddler's utterances—sentences can stretch to six or seven signs, with some rudimentary elements of grammar included.

My host and translator for the encounter with the signing chimpanzees is Dr. Mary Lee Jensvold, a researcher who has worked with Washoe and company for many years and is also fluent in AMESLAN, or American Sign Language. As we watch the languid activities inside the chimpanzees' enclosure, Loulis comes over to the glass partition and sits a couple of feet away, staring at me.

"Hi Loulis," Mary Lee signs.

Loulis answers by looking at me and saying, "Give me an apple."

"He doesn't have an apple," replies Mary Lee.

But Loulis says, somewhat more insistently, "Give me a sandwich."

Then the stereotypical "Give me a banana."

Finally, convinced by Mary Lee that I haven't brought him anything, Loulis gestures to me, "Show me your shoes."

I obligingly lift my foot so he can get a look at black running shoes through the glass partition.

"Black shoes. Nice," Loulis observes. Mary Lee explains that Loulis is fond of the color black. Whether this is connected in his mind with his own color is not clear, but he, and Washoe too, will sometimes thumb through stacks of photos and stop to sign "black, black," when they come across their favorite color.

What does this use of language by chimpanzees tell us about ourselves? For years some linguists and psychologists have denied that language is used by great apes, but gorillas, bonobos, and orangutans have all been shown to be linguistically savvy in laboratory studies. Critics claim that the use of symbolic language by chimpanzees is so limited that we should not call it language. In part they are correct; what chimpanzees do with ges-

173

tural language is equivalent to the speech of a two-year-old child. But I suspect all linguists would agree that two-year-old children possess language, even though they haven't fully mastered it.

Language allows an individual to transmit culture across time and space. Information can be passed from one society to another, and from this generation to the next. The language skills acquired by chimpanzees reared in human home settings are not easily transferred to an ape's natural habitat. In the wild, chimpanzees use a wide range of calls, some of which rely on context to provide meaning. And while the known vocabulary of wild chimpanzees consists of about fifty calls, they use gestures to communicate as often as they use vocal communication.

Do apes have language? First we need to consider what language is. Language is grammatical; it has rules and regularity of use, learned by children without a great deal of hands-on training. It is open, allowing a nearly infinite variety of possible combinations of relatively few sounds in order to convey an enormous array of meanings. And it is symbolic; there is nothing in the word "green" that tells you anything about its meaning. We apply "green" as a label to a well-understood feature or object. Language allows for the indirect transmission of culture across wide geographic areas and across generations of time. The well-documented acquisition of language by chimpanzees reared in human home settings is not comparable to what apes do in the wild. Wild chimpanzee vocalizations rely heavily on context to provide meaning, and their calls do not resemble human language at all.

Our key question about ape language should not be "Do apes have language?" but rather "Do apes and humans share the same foundations for language?" Since the concept of language is a human invention, it is not surprising that linguists and cognitive scientists have been reluctant to apply it more expansively to other linguistically skilled animals. If a chimpanzee shares with a human child the basic "software" for language acquisition (the ape lacks the vocal "hardware" needed to produce intelligible speech sounds), this would be a profound statement about the deep history of language in the early human lineage.

Perhaps the most persuasive research on language acquisition in apes is Sue Savage-Rum-baugh's work with Kanzi, a male bonobo. Kanzi communicates by touching symbols on a lexicon board; he receives information through his understanding of spoken English. Savage-Rumbaugh estimates his production vocabulary as 300 words and his understanding of spoken English as over 1,000 words. Work by Savage-Rumbaugh and many other researchers has conclusively settled at least two arguments over ape language. First, she demonstrates that apes understand and employ the concept of reference, symbolically using words to represent things in their environment. Second, these words and phrases are spontaneously combined to request and give information as well as to comment on or describe the world around them. This is no mere parroting; Kanzi can follow complex requests given in spoken English, such as, "Go to the refrigerator in the next room and bring me the red ball on the top shelf." Changes in word order create the same sort of confusion for a chimpanzee that they would for you, and simple syntax is present and necessary. If there is a difference between what Kanzi comprehends and what a human toddler comprehends, scientists have not yet discovered it.

Above all else, language is a social behavior that enables higher-order communication among individuals. Like many other chimpanzee and human social behaviors, it is learned through socialization. After the Gardners' and Fouts' work with signing chimpanzees, a psychologist named Herbert Terrace attempted to replicate the result with a chimpanzee he nicknamed "Nim Chimpsky" in mock honor of the famed linguist. Terrace made one key mistake, however, in preparing Nim for language research. Instead of rearing his study subject in a social environment in which the intense mother-infant bond was approximated (as it was when Washoe spent her early months and years raised as a human infant), Nim was raised in a laboratory. Instead of one or two committed surrogate parents, Nim had a rotating host of students and lab assistants taking care of him. And by all accounts his social environment in those early formative months was impoverished compared to what Washoe had received. And that may explain why Washoe was a skilled language-user, who in turn taught sign language to her own chimpanzee peers and adopted son, while Nim turned out to be an underachiever. Terrace concluded that Nim's language use, more limited than Washoe's, was mere clever imitation rather than linguistic proficiency, and the results cast a shadow on ape language work for years.

In a laboratory in Japan, a chimpanzee named Ai displays math conceptual skills that would make any teacher proud. She has learned to recognize that Arabic numerals stand for particular numbers of objects. She has the ability to count from zero to nine, which she does by pressing a touch-sensitive computer screen. She can properly order the sequence of numbers from zero to nine when they are flashed to her on the screen so quickly that a human would be hard-pressed to compete. The average person can remember numerical sequences easily to about seven digits—think of phone numbers—but after that our ability to remember such information rapidly diminishes. Ai is proficient at recalling number sequences up to five digits, about the same number of digits a four-year-old child can remember. She can even interrupt her sequence to attend to something else, like food or a fight in the group, and then return and complete the sequence. This basic math competence in an ape was shocking when first reported by the Japanese researcher Tetsuro Matsuzawa.

Mirror, Mirror . . .

To be capable of distinguishing oneself from others or to somehow recognize oneself as a distinct entity is another attribute of what we call cognition. When I wake up in the morning and look at my face in the mirror, I recognize the image as myself. Consequently, I've demonstrated a sense of self-awareness. When my dog Burbank walks in front of the mirror, he sometimes stops briefly to face himself, but he probably doesn't recognize that image as himself. The outcome doesn't change if I place a hat on his head; Burbank's perception of the image remains exactly the same.

But is this true of all animals other than humans? Proving the existence of a "sense of self" is experimentally challenging, but a few successful experiments that use mirrors and markings as tools have been carried out on various species in recent decades. The experiments were simple. Mark the body of the subject animal with a splash of paint or dye, encourage the animal to look at itself in a mirror, then observe and record its reactions.

Chimps, gorillas, bonobos, and orangutans blotched with paint and stationed in front of a mirror were recorded examining and touching the actual spots on their bodies. They were not only able to recognize themselves in a mirrored image but were also capable of understanding that the blotches of paint had occurred on their own bodies, not on that of the animals in the mirror. Recently, one female Asian elephant named *Happy* was also able to pass the mirror self-recognition test, repeatedly touching a white X on the side of her head with her trunk.

Similar experiments with dolphins were complicated by their lack of arms and hands. Dolphins were marked with black ink in an area of their bodies not visible to them. They could, however, feel the ink. A mirror was then offered, and the dolphins were watched to see if they were visually monitoring their bodies to find the ink spot. Some marine mammals, including bottlenose dolphins and killer whales (and perhaps false killer whales), showed a positive self-recognition response in that they were observed visually inspecting themselves in the mirror to find the marks.

These experiments on self-recognition and self-awareness in dolphins do not, by themselves, prove that these animals are self-aware in human terms. Scientists are still at odds over the degree to which dolphins are capable of self-consciousness, with some arguing that the animals do not have a concept of self and are not cognitively advanced at all. Proving that dolphins and great apes are able to recognize themselves in a mirror, however, is an important step toward establishing these animals as cognitive species and toward suggesting convergent cognitive evolution in our next of kin.

Great apes and dolphins might not think like humans, but researchers have collectively established that they are able to memorize, imitate, use tools, understand language, and be self-aware, even if we are still squabbling over the exact definitions and the degree to which they are capable of these things. We have, in the preceding chapters, laid the foundation to envisage the many parallels between our intelligent world and theirs, but things become even more interesting when we begin to look at what Italians call the *sfumature,* what is beneath the words. Politics and culture are things thought to be exclusively human, but are they?

chapter six

MASTER POLITICIANS

GREAT APES POSSESS an intellect that is often referred to as Machiavellian; they remember favors owed and debts incurred, and operate a service economy of behavior exchange. Male chimpanzees form paramilitary patrol parties and hunting parties. Individual male chimps engage in manipulation and deceit of rivals to Shakespearean proportions.

Some dolphins make use of their own so-cial intelligence to form coalitions of males that sexually coerce females and overthrow other male coalitions. These coalitions evolve into alliances— highly complex behavioral "agreements" between males of the same school that cooperate in pairs and triplets to sequester and control the move-ment of females likely to be in estrus. These sorts of multiple-level alliances within a social group re-quire a keen mind of a sort that is not seen in other animals. But coalitions are only one aspect of these animals' political repertoire. Not so unlike human politicians, dolphins can also be masters of deception and manipulation, with sharply honed networking skills.

To illustrate how essential political savvy is to these creatures, we must navigate their social worlds to explore, with different examples from the oceans and the forests, this unusual part of their existence.

The Art of Deception

In our human society, intentional deception is commonplace. We deal with it every day of our lives. Politicians deceive us. Salespeople trick us into purchases that we might not otherwise make. We must daily negotiate a myriad of manipulated information, misrepresentations, fabrications, and spin. Deception with forethought requires a large brain, cognitive skills, awareness, and an ability to think on the fly. It requires a theory of mind. But are we the only species on the planet capable of intentional deception? Observations show that there are others whose swindling skills rival our own. Large-brained animals like dolphins are particularly good at deception and, in some cases, seem to do it with some degree of intention.

Stan Kuczaj has worked with captive dolphins for many years. He and his colleagues have recorded several occurrences of what could be considered deceptive behavior in dolphins. At one of their facilities, the subject of their research was a female adult dolphin named Kelly, who along with her tankmates had been trained quite successfully to retrieve objects from the pool in exchange for fish. After all the other dolphins had finished with their retrieval chores and gone their own way, Kelly appeared at the surface with some objects of unknown origin in the hope of gaining more fish.

Where did Kelly's objects come from? After searching the pool, Kelly's trainer discovered a secret cache of toys that the dolphin had astutely concealed under a drain cover. Day after day, she had collected objects inadvertently dropped into the pool by tourists, to be used for barter with her trainers for fish. On closer observation, it became clear that Kelly was extremely careful not to add or remove objects from her cache when other dolphins were present. Was she intentionally deceiving other dolphins by secretly hoarding these objects for her own use or was this just a coincidence?

Another interesting case was that of what we might call the "paper dolphin." This female, like Kelly, was trained to bring objects to her trainer in exchange for a reward. In one session, at the trainer's request, she arrived at the surface with a small piece of paper picked up from the bottom of the pool ready to claim her reward. Shortly thereafter, she brought another piece of paper to exchange for a second fish. Then a third, a fourth, and so on; always small pieces and always one at a time. After a sizeable pile of paper bits had amassed near the poolside, her trainer decided to make a closer inspection of the pool. Looking carefully, he discovered what remained of a large paper bag stuck in one of the underwater grates and subsequently gave the dolphin the command to retrieve it, which she did in its entirety. It was unclear how the bag became trapped in the grate and whether the dolphin had anything to do with its being there, but she had succeeded in fooling her trainer and increasing the rewards received by retrieving bits of the paper bag instead of the whole object.

Deceptive behavior by wild dolphins is also found in the annals of scientific literature. Janet Mann of Georgetown University observed a band of female bottlenose dolphins using deception and misrepresentation to hoodwink an alliance of aggressive males attempting to harass a single outnumbered female in estrus. Swimming alone, the female had been completely surrounded by the group of males when a band of females arrived to her rescue, diverting the males' attention by rubbing up against them and stroking them with their fins. Once they had succeeded in confusing and misleading the male alliance from their initial target, the female group, with the rescued female among them, moved quickly to safety.

Dolphins are not the only large-brained mammals other than humans that hoodwink others. Both monkeys and apes have shown skill with these tactics. Dorothy Cheney and Robert Seyfarth, who studied vervet monkey social behavior in Kenya, wanted to know the extent to which the monkeys understand the nature of social relationships within their group, and how that knowledge might be used by group members to manipulate and deceive one another. Vervet monkeys are clearly good anthropologists; they understand kinship patterns in the group. Using tape-recorded playbacks of baby vervet distress calls issuing from cleverly placed loudspeakers, Cheney and Seyfarth showed that vervets are keenly aware of which baby belongs to which mother. When the call of a given baby was played by the researchers, the other animals in the group looked at the mother of the missing infant.

Female vervets, in other words, understand patterns of maternity in their groups.

The researchers also found that vervet monkeys lie. Many primatologists have seen examples of lying in primates. Lying appears to be an evolutionary trend, becoming more widespread in the bigger-brained higher primates. For example, great apes seem to be skilled at deceiving one another, whereas lemurs rarely if ever do it. In the vervet study, one monkey would give a predator alarm call as the group fed in a desired fruit tree. As other group members fled from the "predator," the call-giver would capitalize on its lie by feeding aggressively in their absence. I once watched a low-ranking male chimpanzee named Beethoven mate with a female despite the presence of the alpha male Wilkie by using deception. As a party of chimpanzees sat in a forest clearing, Beethoven made a charging display through the middle of the group. Because he was a low-ranking male, this was taken by the alpha Wilkie as an act of insubordination. As Beethoven charged past Wilkie and into dense thickets, Wilkie pursued and launched into his own display. While Wilkie was absorbed in his display of dominance, Beethoven furtively made his way back to the clearing and mated.

In the worlds of dolphins and primates we find examples supporting the idea that these animals may intentionally deceive others to gain a variety of benefits. Deception with forethought requires a complex brain, cognitive skills, awareness, and ability to learn new solutions to problems. Dolphins and apes seem to possess all of these traits, at least in some measure. Intentional deception is not solely a human talent, but one that also belongs to these animals of the forest and the ocean.

A Service Economy

In the ape's world, as in ours, size does not matter as much as smarts. U.S. politicians do not become president because they are tall, and male chimpanzees do not ascend to alpha rank by being physically big or brutish. An astute mind and a clever sense of the political calculus better serves the politician, who must persuade people that he or she is the leader they want. In the same way, a male chimpanzee rises in rank by currying favor with the right group members; if he learns how to make use of alliances with powerful group mates, both male and female, he will go far. At Gombe, many of the male chimps that are the largest in size do not even approach alpha rank; in fact there is some tendency for smaller males to end up dominating larger ones. But the one thing nearly all alphas have in common is a shrewd sense of favoritism, timing, and alliances.

How, you may ask, can a nonverbal animal be such a clever politician? Without words, chimpanzees and other great apes engage in a complex service economy, in which favors owed and debts to be repaid are a currency for securing needed social support. Other exchanges of services may occur too. This is not novel to primates; many animals engage in a market economy of sorts, in which behaviors or resources are exchanged. Animals in a position of power—such as high-ranking chimpanzees—may not need to engage in a fair exchange.

A low-ranking male needs the support of a high-ranking male. Therefore, the higher-ranking animal can get away with offering his support, in a fight for example, on only rare occasions, because that is still more charity than the lowly animal can expect elsewhere. Such an imbalance of power is often a defining feature of the social networks found in many social animal species, especially the more cognitively complex ones.

Frans de Waal showed that chimpanzees may engage in a service economy in which food and grooming are currencies swapped for each other. Grooming in chimpanzee society has multiple functions; it is communication, analogous to our small talk; a way of bonding with others; and, of course, a convenient way to rid oneself of parasites. Chimpanzees who receive food from other group members are more likely to groom those donors. Food sharing is not common among chimpanzees, except between mothers and their infants. When a chimpanzee willingly passes a morsel of food to another, it is done for a reason. The main reason is kinship. But chimpanzees do share with others at times, usually in strategic ways. For instance, following a successful hunt, male chimpanzees share meat from the kill with others present. Sharing is not random or liberal. Captors share with allies, and snub rivals. They readily share with their mothers, sisters, and brothers. They eagerly share with females who possess sexual swellings, or who are desired future mates.

After one particularly successful hunt of colobus monkeys in which seven monkeys were caught, the male chimpanzees in the hunting party sat for hours feasting on their treats. Females and young or low-ranking males approached meat holders and begged, using the very human gesture of extending the hand in a supplicatory way. When this didn't work, the beggar actually placed his hand at the lips of the meat-eater in hopes of appealing to his charity, or perhaps just to persuade him that surrendering a bit of cherished meat was better than fending off hours of demands. In another case, adult male Atlas captured a young colobus monkey that virtually leapt into his arms as it tried to flee during the hunt. Atlas had bitten the monkey at the base of its skull, and the small carcass hung limply in his hands. Moments after the capture, an adult female named Trezia rushed to the scene, her hand extended in a begging gesture. Apollo, Atlas's little brother, also arrived. As Trezia approached Atlas, a smile-like fear-grimace stretched across her face, he turned toward her but held the body of the monkey behind him, at arm's length. He continued to do this until Trezia turned and presented her swelling to him. While they mated he continued to hold the colobus out of her reach. After several mating bouts he offered a share of the meat to her and also to his little brother Apollo.

Although meat is often used as a bargaining chip to cement alliances with males and desired females, male chimpanzees hunt avidly even when no females are present, and meat and sex often occur in the aftermath of a hunt without such an obvious barter occurring. I am often asked if meat-sharing in exchange for service rewards represents the origins of a service economy or even the origins of prostitution. The best way to view male chimpanzee use of meat is in a wider arena of social manipulation. Monkey meat is valuable to females because it provides a dense package of protein, saturated fat, and calories, all essential during times of gestation and lactation. But female chimpanzees rarely hunt, and when they do they are rarely successful. So meat remains a highly limited and sought-after commodity. Female chimpanzees are, as we have seen, highly promiscuous and strategic when it comes to mating, pairing briefly with many males during the course of a single ovulatory period. This likely confuses the issue of paternity enough that each adult male has some disincentive from harming an infant who may be his offspring, and perhaps also discourages him from unduly harassing the infant's mother.

Being a skilled hunter is one way to get monkey meat, but being high-ranking or the kin or ally of someone high-ranking is another option. After a hunt, a high-ranking male will often steal colobus carcasses with impunity from low-ranking or immature males, or from the occasional female who makes her own kill. The alpha male thus uses his status as a means to control access to a limited resource that is desired by all, and he can dole out to his allies and withhold from his rivals. In this, he is very much the politician, securing "votes" in the form of favors in exchange for the spoils of his power.

This sort of political behavior appears to be

widespread in chimpanzees and may reflect their common ancestry with us. When Newt Gingrich became the Speaker of the House in U.S. Congress in the 1990s, he publicly mentioned Frans de Waal's book *Chimpanzee Politics* as an influence on his ability to manipulate those around him. The political machinations of humans also tend to be related to the acquisition of power or other resources.

In chimpanzee society, it is the males that engage in Machiavellian behavior. Researchers used to think that the gregariousness of male chimpanzees was based largely on kinship. Since males do not emigrate from their home community, it would make sense if most of the males in a community were relatives. Indeed, during my own research among the Gombe chimpanzees, the eleven adult and adolescent males consisted of four pairs of brothers (Frodo and Freud, Atlas and Apollo, Prof and Pax, and Goblin and Gimble). An eleventh male, Beethoven, had arrived as an orphan in the company of his apparent older sister. But male kinship within a community is not the norm among chimpanzee populations. In studies at other chimpanzee study sites, male kinship has been found to be low. The males at these sites work together to sexually coerce females, gang up on high-ranking males, patrol territories, and hunt based on shared interests, not the bond of genes.

This is an interesting and surprising finding. It tells us that, like humans, chimpanzees make strategic and tactical decisions about how to behave in their social lives based on rapidly changing events. A male chimpanzee will flip-flop his allegiance from day to day, depending on how the situation of the moment fits his agenda in life. In this sense, male chimpanzees have political careers, in which the goals stay more or less the same—wield as much power, influence and reproductive success as possible—but the tactics for achieving them vary from day to day, year to year, and life stage to life stage.

Male cooperation in wild chimpanzees does not seem to be based on genetic kinship. Nevertheless, males spend their lives trying to rise in rank toward alpha status. Why? Researchers had always assumed that the goal and the benefit of becoming alpha was the opportunity to father many baby chimpanzees. But early genetic studies seemed to suggest nothing of the sort. They showed instead that alphas did not father more infants than, say, a sneaky adolescent. A study in the 1990s in western Africa seemed to show that female chimpanzees in the study community were actually being inseminated by males from other communities. The authors concluded that far from monopolizing mating success with their own females, males were being duped into protecting females who were shopping for genes with males in rival territories. The general promiscuity of male and female chimpanzees also cast doubt on whether alpha males really reaped a reproductive benefit from their status.

All this seemed to suggest that the benefits of high rank must be something other than reproductive success. Perhaps there are psychological benefits—studies of human males have shown that high-ranking CEOs have lower levels of stress hormones than mid-level underlings. But in Darwinian terms, such psychological effects don't count for much unless they translate into leaving more offspring.

But then new research clarified the relationship between male rank and reproduction. The earlier finding that half of all babies were fathered by males from outside the community was invalidated when some methodological problems with the data analysis were uncovered. And researchers in Gombe National Park, the longest-running study of wild chimpanzees—completed their own analyses on paternity and found that the alpha male fathered the vast majority of offspring. So the benefit to spending one's life trying to climb the dominance hierarchy was a reproductive payoff after all.

How males achieve reproductive success is still unclear, however. Remember that the alpha male at his physical prime and a beautiful creature at age twenty-five carries the same genes that he carried as a runty teenager or will carry as a grizzled old codger. We know that females select males based on some aspect of their phenotype—their outward appearance—but in doing so they receive a genotype's worth of DNA. So the female who sneaks off into the thicket with Frodo when he is a ten-year-old, risking the wrath of the alpha, may also mate with him when he is the alpha himself a decade later, or when he is a frail creature in his late forties. All three versions of Frodo pass on the same good genes.

Female chimpanzees also align themselves in a dominance hierarchy, albeit a less apparent one than we see among males. There is a matriarch of sorts, with whom males on the rise curry favor in hopes of securing her support in contests over rank. Anne Pusey and her colleagues found that at Gombe National Park in Tanzania, female rank was tied to female reproductive success; the high-ranking female, with the exception of one, tended to give birth to more offspring.

Since females are strategically promiscuous but also seek to find good genes for their offspring, how do they choose a mate? They seem to select them on the basis of traits that are not always apparent to observers. Apart from rank, neither physical stature, musculature, age, or appearance seem to strongly predict which male will be a female's preferred sexual partner. Males choose mates too, and their standards of beauty are not obvious to human observers. Flo was the famed matriarch of Goodall's early years at Gombe; she was the highest-ranking of the female chimpanzees and highly sought after as a mate. But she was not beautiful in any human sense of the term; she had a ragged ear, a scarred face with a bulbous nose, and by her forties, a somewhat gaunt body. But she was wildly popular with the males. Presumably, something about Flo, perhaps her high social standing or even something olfactory that was beyond the ken of human observers, brought her males throughout her long life.

The Deal

In their business book *Strategy of the Dolphin: Scoring a Win in a Chaotic World,* Dudley Lynch and Paul Kordis propose the dolphin as the exemplary model of the "ideal negotiator" because of its large brain and its ability to remember and learn from experience. Perhaps the concept of dolphins as negotiators might be pushing the envelope, but there is probably some truth to what the authors purport.

When a dolphin doesn't get what it wants, it cleverly modifies its behavior to obtain what it seeks, much like a human child who may cry and scream to get an ice cream when the parents refuse to buy it. Sometimes, if crying and screaming don't work, a child may try a new ploy like nagging to gain success. If the new approach works, the child will remember it for future exploitation, and so it seems for some dolphins as well. Because life in the ocean means coping with patchily distributed resources that may be abundant at times and scarce at others, dolphins must learn to adapt to the different situations they encounter. Finding allies, building trust, and getting along with peers can be a winning strategy for a variety of benefits.

Transient and resident killer whales use a type of "agreement" to co-exist in the same environment: they stay socially isolated from each other and feed on entirely different prey. Aggression between transients and residents is rare, which tends to prove that this agreement in a sympatric environment, that is, one in which two species coexist but do not breed, works quite well to maintain stability in the community.

When talking about dolphins, deals can break species boundaries at times. In *Natural History*, Pliny the Elder was the first to describe a sort of "agreement" between wild dolphins and humans.

He narrated the association between dolphins and fishermen in the French village of Lattes, in Narbonne province. At one time of the year, as the tide receded, huge schools of mullet left the pond in Lattes to reach the open sea, passing through a narrow waterway. Because the fishermen could not use their nets in the narrow passage to the open sea, they called for Simo the dolphin. Every year Simo came to aid the humans in their attempt to catch mullet. When the fishermen called, Simo assembled a troop of dolphins as a general might assemble his soldiers before a battle. The dolphins moved in a rank near the narrow exit, assaulting the huge schools of mullet escaping through the channel and driving them back to the nets and pitchforks of the fisherman, away from the safe waters of the open sea. After sharing the catch with the fishermen, the dolphins would return to the sea and the fishermen to their homes with enough fish to feed their families. Might this narrative be just a fable? We do know that this type of dolphin behavior also exists today, and accounts of dolphins and fishermen collaborating abound in different parts of the world.

In his book *Dolphins,* the explorer, filmmaker, and oceanographer Jacques-Yves Cousteau tells of the local Imragen fishermen of Mauritania and the dolphins that help them fish. Fascinated and intrigued by this alleged collaboration, Cousteau sent a team of divers and cameramen to document this man/dolphin cooperative effort. As schools of mullet migrated along a stretch of African coastline, dolphins would aid the Imragen by circling and herding the fish into the nets of the waiting fishermen. Cousteau's cameramen captured this remarkable collaboration on film, bringing to reality what had previously been only a story handed down from generation to generation. There are no stipulations or signatures in such interspecies "agreements," no formality or form. There are only two species, living in different worlds, that seem to work together for a common benefit.

Networking at Sea

A network connects things together. When we think about a network, we may imagine a series of computers linked to share information. But human society is another great example of a network. Our social network is made of nodes, either individuals or organizations within the network, and ties, the relationships between them. Networks can be more or less open, with strong or weak connections between the members. Network theory explains how social networks can function at different levels, from families to entire nations. It plays an important role in how individuals exchange information, attain goals, and solve problems.

If people have social networks, why shouldn't dolphins have them as well, considering the complexity of their societies? Like us, some dolphins form strong and weak relationships and alliances and are able to access and quickly exchange a wide range of information. Thinking about networks and dolphins, David Lusseau and Mark Newman tried to identify the role that dolphins play in a social network in the waters of Doubtful Sound, New Zealand. Seeking to understand underwater networks is not an easy task. As the researchers state, "dolphins can't grant personal interviews or fill out questionnaires." The only way to understand their social network is to study their behavior directly. So, from 1994 to 2001, these scientists focused on a small community of sixty-two dolphins in an attempt to untangle the nodes (dolphins) and ties (associations between dolphin pairs), shedding light on the roles played by different individuals in maintaining the network.

What they discovered was that within the same population, there are communities and sub-communities that are sex- and age-related. There are also centralized "brokers," usually adult females that function as links between these sub-communities, helping the entire network to remain together. By watching these brokers, the researchers were able to gain a sense of how information flows through the network.

In a computer network, the removal of a broker or an important node would result in a shutdown or impairment of the network function. In the dolphin world, however, Lusseau and Newman discovered that the removal of a broker may not impact network function at all, illustrating that dolphin communities have a high degree of flexibility and cohesiveness.

The Politics of War

Male chimpanzees are fierce rivals at times, battling and even killing each other over mating rights, territory, and other valuable commodities. But when it serves them, the very same chimpanzees can become tight allies. They form coalitions to coerce and control females, to patrol their territorial boundaries, and to hunt for monkeys, pigs, and other meaty prey.

Some dolphins do exactly the same things. Male bottlenose dolphins in the warm waters off western Australia form long-lasting alliances. They use these strategic friendships to maraud other male alliances, steal females, and coercively kidnap them for their own sexual purposes. Male coalitions may remain together for years, but others are brief and serve only a short-term strategic purpose.

These coalitions also serve both bottlenose dolphins and chimpanzees well in hostile encounters with other groups. When Jane Goodall first saw commando raids, which she dubbed "warfare" between neighboring chimpanzee communities, skeptical researchers doubted the claim and later suggested the behavior was pathological. We know today that male chimpanzees all across their range in Africa patrol their territorial borders and fight viciously to maintain them in the face of intrusions by neighbors.

To be on a patrol is exciting, even when the chimps don't hear or see any intruders. The atmosphere is tense and quiet, because the males are afraid of being ambushed by a patrol party from enemy territory. A normal chimpanzee party on the move features a lot of play, roughhousing, sex, squabbling, and munching food as they go. But a patrol is obvious to anyone who has seen one. The male chimps glance about nervously, maintain a code of silence, and sniff the ground and objects around them. This is serious stuff, not unlike the human counterpart of a military patrol. Although some descriptions of chimpanzee patrols give the impression that the males are hoping to meet and attack their neighbors, nothing could be further from the truth.

Just as in a military platoon on patrol, the goal

is to find stragglers, members of the enemy community who were out traveling, got a bit lost, and ended up in the wrong place at the wrong time. When a patrol meets a lone chimp from another community the enemy usually does not survive the encounter without serious injury or worse. The ambushed chimp screams in mortal fear as it is set upon by the patrol, which use teeth, nails, and arms to deliver punishing blows. Even if the stranger is female, she may be brutally attacked unless she is young and childless and displays a sexual swelling. In that case the patrol will coerce her to follow them back to their community rather than subject her to aggression.

Bottlenose dolphins utilize "sentinels," individuals stationed strategically apart from the group. These stand guard while the other members of the school are feeding. What may appear as altruistic behavior on the part of the sentinels might be better interpreted as a benefit to the individuals—they are less likely to be caught off guard themselves if they are on patrol. Once an uninvited guest or a predator has been spotted, the dolphins may react in a variety of coordinated ways; they may escape, they may monitor the predator with a careful inspection, or they may mob the trespasser, harassing and attacking it.

Like chimps, dolphins use body language to scare an intruder. An array of postures and sounds are used to express different degrees of aggression. When bottlenose adult males encounter an intruder, they may arch their back in a coiled, aggressive posture. In response, the rival may ready for a conflict by opening his mouth and displaying his sharp, white teeth. Should the encounter escalate to a conflict, jaw claps, biting, and a chop inflicted with a flipper may invoke a semi-comical squawk from the intruder, who will generally retreat by swimming quickly to a safe distance.

This chapter ends on a note of war, which one might argue is the distillation of political ability. Political acumen is that which gives us the ability to manipulate power and control within our societies. We have shown that great apes and dolphins also possess elements of political savvy.

Political ability might be seen as one expression of human intelligence and accomplishment in the universe, and culture is arguably another. If political saavy is not entirely restricted to humans, is culture? In the next chapter, we will endeavor to show that we are not the only ones capable of cultural facility.

chapter seven

CULTURE VULTURES

ALTHOUGH WE, the authors of this book, are both biological organisms—primates, in fact—we define ourselves by our cultures. The two of us speak English (one accented in Italian), wear western clothes, eat mostly western food. We are not part of a caste system, we do not live in mud huts, nor do we have animistic religious beliefs.

The word "culture" is derived from the same Latin root as "cultivate," suggesting a connection to how the human mind and behavior are nurtured and developed. To many, culture refers to things like going to the opera or being able to taste the difference between a Sauvignon and a Chardonnay. But culture is a far more basic trait. It refers to the sum total of the learned, traditional aspects of a species; beliefs, institutions, cuisine, ways of dress, and all other products of a group. Since a group can often be characterized by all of these expressions of itself, we call that a culture. That sounds simple—anything learned is culture, whereas anything inherited is genetic. But the issue is far from simple.

Culture is a profound enigma, a nearly amorphous, slippery concept whose identity lies in the mind of the beholder. It has spawned whole intellectual fields, whose devotees argue endlessly and futilely about what culture is, what it should be, and how much control it exerts on the human spirit. This last issue is a thorny one. Are humans cultural creatures only barely constrained by an ancient, now largely irrelevant biology? Or are we biological creatures, our urges and impulses filtered through a thin film of culture? Clearly the truth lies somewhere in the broad middle, and that is a heavily fortified intellectual battlefield where wars are waged among researchers.

Anthropologists are generally accepted as the experts when it comes to defining culture, but even among them there is a deep split. Cultural anthropologists, who study human societies, consider symbolism and symbolic activity to lie at the heart of any definition of culture. Humans are human, such anthropologists say, because they arbitrarily attach names to things without regard to their literal descriptive power—"red" does not connote anything about the color red at all—they are learned symbols. Many biological anthropologists and those who study animal behavior have adopted a broader definition of culture.

A key aspect of this debate over culture and human nature involves whether the concept of culture can or should be extended to other animals at all. Ever since Jane Goodall observed tool making by chimpanzees, these apes have been regarded as the "cultural animals" that blur the distinction between human and nonhuman. Cultural traditions are those behaviors that are learned from one's peers and family, and are passed down through generations by repeated learning, not through genetic inheritance. By this definition, cultural behaviors can include how tools are made and used, and also the style in which one chimpanzee grooms another.

Let's return to the termite mound mentioned in Chapter 5, for example. In a forest in East Africa, a chimpanzee approaches a large, conical termite mound. She reaches out to a nearby bush and plucks a leafy twig, then pulls the twig through her lips to pull off the leaves. She carefully grasps the bare twig and strides to the termite mound. Inside this concrete-hard earthen hill lives a colony of nearly 10 million termites. These are not the diminutive termites seen scaling rafters in your garage. The soldiers that protect the colony are large—a half-inch long or more—and armed with a formidable set of mandibles. They serve the colony by keeping intruders, especially intruding termites, from entering the many tunnels that connect the colony deep underground with the outside world.

In East Africa, the start of the rainy season in October or November brings the termites nearer the surface as the annual reproductive cycle of the colony peaks and the future queens crawl out of the tunnels, their new wings still flattened to their bodies. These youngsters then spread their wings and take flight, creating clouds of termites across the landscape and scads of landed queens crawling underfoot on every footpath and in every local home, where they are gathered up and eaten by local people, who appreciate their nutty flavor. As the rains intensify, the clay soil of termites' massive mounds softens, and the chimpanzees reap the bounty. A termite-fishing chimpanzee first quickly inspects the mound's surface. Then she uses her index fingernail to pick at the opening of a promising tunnel. Once opened, she inserts her fishing probe, holds it there for just a few seconds, and gently withdraws it. Typically, the end of the probe is carpeted with chocolate-brown termites, both tiny workers and chunky soldiers clinging by their mandibles. She runs the tool quickly through her lips, and crunches a snack of protein and carbohydrate-rich insects. If the mound is productive, as most are during the rainy season, she may stay for hours and reap thousands of termites. Her offspring may sit by her side, watching her technique intently and attempting to imitate her.

Chimpanzee tool use varies widely across Africa in a pattern that suggests that appearances and extinctions of local tool technologies may occur frequently. Despite much interest in the roots and the spread of tool use, there is no evidence that the use of particular tools is related to the particulars of a given forest; one chimpanzee may termite-fish in forests where termites aren't abundant, and another may ignore them in forests where termite mounds are a dime a dozen. There is also no evidence that genetic differences account for different patterns of tool use among chimpanzee populations. Instead, the local cultures of chimpanzee tool use appear to reflect traditions that arise in individuals and spread within, and perhaps among, breeding populations.

Researchers have made some generalizations about tool use. Chimpanzee tools fall into one of three types. They use stick tools to increase the reach of their arms ("wands" for dipping into driver ant nests or probes for obtaining honey) or maximize their arm strength (wielding a stick against a rival). They use "sponges" made of chewed leaves to absorb rain water from tree cavities. And in some parts of Africa, such as Taï National Park in Ivory Coast, they use hammers of wood or stone to crack open nuts and other hard-shelled foods. These traditions themselves vary among nearby communities. Gombe chimpanzees fish for termites, but in Mahale National Park, only a hundred kilometers away, chimpanzees fish for ants from tree trunks using the same methods but never apply this tool to termites, even though they are abundantly available.

When it comes to ape tools, availability is not the mother of invention. Gombe possesses an abundance of rocks of all sizes and shapes and widespread nuts and fruits. But Gombe chimpanzees have never been reported to use stones as hammers. Meanwhile, in the Taï forest, rocks are few and far between compared to Gombe, yet the Taï chimpanzees forage for them and carry them back to nut trees to be employed as tools. The pattern of tool use that we see at each site is nearly the opposite of what we would expect if tool use were based primarily on the availability of particular forest resources.

The problem with the study of chimpanzee technological culture has been that it consists mostly of anecdotes, which are fascinating but don't convince other scientists. When chimpanzee researchers had finally obtained enough long-term data to systematically analyze cultural traditions from a range of field sites, however, they found unequivocal evidence for a systematic pattern of cultural traditions. Scrutinizing tool use and other cultural data from the seven longest-running field studies in Africa, Andrew Whiten and his colleagues found at least thirty-nine behaviors that could be attributable to the influence of learned traditions. This may seem rather limited compared with the human catalogue of learned behavior, but compared with other nonhuman animals, it is an impressive statement. The logical conclusion might be this: animals that live by their wits, as it were, tend to be like chimpanzees and humans, benefiting from big brains and long life spans. During growth and maturation, key life skills are acquired by watching one's elders and peers.

The bonobo, with its rich repertoire of social and sexual behaviors, is a Luddite when it comes to technology. Although much attention has been given to the sophistication of bonobo social dynamics, sexuality, and male-female relationships, bonobos are poor tool users in the wild. They are known to drag branches across the forest floor as a means of signaling others in the foraging party of their desire to travel. But there are no reports of the use of either stick or stone tools in any of the contexts in which chimpanzee tool use is commonplace.

Gorillas, too, are very limited technologically. Although these largest of the primates sometimes eat termites and ants, they obtain them with their fingers, and no reports of tool use to insect-fish exist. Researchers recently made a stir when they reported the first tool use by wild gorillas. A lowland gorilla was seen to use a large branch as a probe to gauge water depth as it walked upright across a swampy clearing. As you might poke a stick in front of you when crossing a flooded street, so the gorilla walked along, used the stick to find bottom as he walked. This is very impressive for a gorilla, but it pales in comparison to many everyday sorts of tool use by chimpanzees.

In the 1990s, orangutans joined the list of tool-using primates when Carel van Schaik observed these gigantic red apes making and using very simple tools in Sumatran forests. Orangutans live in a complex world of high tree limbs, tangled vines, and lianas, and the tools they make are primarily probes used to extract small animals and plants from hard-to-reach holes in trees. Van Schaik and his colleagues found that across the range of the orangutans in Borneo and Sumatra, populations differed in a variety of behaviors that suggest the same sorts of cultural variation seen in chimpanzees, albeit at a simpler level. Tool use differs from forest to forest. In some forests orangs use leaves as "gloves" for protection when feeding on thistly plants, whereas in others they forage for the same plants bare-handed. Some orangutan populations use leaves or their hands to utter squeaky vocalizations, whereas others do not. And females in some populations even employ sticks as auto-erotic tools for sexual stimulation, whereas other (perhaps more repressed) populations do not. The variety and complexity of orangutan culture may pale in comparison to that seen in chimpanzees, but it is far more frequent and varied than researchers could have imagined just a decade ago and provides evidence that culture is, broadly speaking, widespread among apes.

Tool use in dolphins is much less documented. Studying animals in a watery environment is difficult enough, and only recently have some cetologists embraced even the possibility of tool use in the ocean world. Scientists in Shark Bay, Australia, documented the first case of material cultural transmission in the ocean environment, the sponge-carrying behavior of dolphins discussed in Chapter 5. The use of this effective tool is passed from dolphin mothers to their daughters and has been observed only in Shark Bay. Tool use in cetaceans may be more widespread, but only more time in the field and advances in technology that allow us to see more easily into the underwater world of these animals will enable us to know for sure.

Ape Culture?

The imitative abilities of great apes can be best understood through the lens of culture. Great ape cognition is highly social in nature; the context in which their cognitive skills evolved was, like our own, a complex web of interindividual interactions. Learning through observation and interaction in a social group, whether the subjects be human or great ape, is far different from learning in isolation. This makes the study of learned behaviors under captive laboratory conditions, including the study of imitation, problematic to conduct and the results difficult to interpret. In particular, laboratory studies that rear chimpanzees like lab rats and ignore the importance of stable, long-lasting kin bonds and the social influences of a kin group end up with ape study subjects no more "normal" than children in a war orphanage.

In the wild, the importance of socialization to chimpanzee culture is seen at its best. Chimpanzee cultural diversity is not limited to tool technology. Just as Texans and New Yorkers speak with accents that are jarringly different to a native English speaker, chimpanzee populations may speak with accents, too. John Mitani and his colleagues conducted acoustical analyses of chimpanzee pant-hoots from two forests about sixty miles apart in Tanzania. They found that elements of the pant-hoot showed different acoustical properties—some elements short and others long—in the two sites. They suggested that dialects may exist among chimpanzee populations, just as they do among killer whales. These would presumably emerge as small, random differences in each chimpanzee culture and become institutionalized when youngsters imitate the calls they hear their older peers uttering. Linguists, accustomed through their training to consider only human language to be true language, have a lot of trouble accepting communication like the pant-hoot as language. The academic field of linguistics, like history or psychology, was founded as a uniquely human discipline, and academic linguists have been trained to understand language as a uniquely human attribute and are no more willing to expand their definitions than historians of the African continent would expand the study of history to include the history of the Gombe chimpanzees.

Other social behaviors in apes show regional variation, too, and amount to simple forms of cultural diversity. For example, chimpanzees love to spend hours grooming one another, running their long, slender fingers through a group-mate's hair. In Gombe National Park, Tanzania, chimpanzees sit face to face, grooming each other's bodies. Each groomer braces himself with one arm against a tree limb overhead while the other hand picks through his partner's hair. This is the cultural norm, the accepted community-wide posture to adopt while grooming a neighbor in Gombe. It is like a human society in which people wave hello rather than raise their hands together in greeting, as Hindus in India would. But only a scant hundred miles from Gombe in Mahale National Park, chimpanzees exhibit a different style. A Mahale chimp also sits face to face when grooming, but clasps one hand in the arm of his grooming mate while the free hand looks for parasites in his partner's hair.

This may seem trivial, but it is no more trivial than the myriad behaviors that make up human cultures. There is no reason to think that genes or environment dictate this difference in grooming posture; it is simply a cultural tradition. Scientists have long been fascinated by how such traditions migrate from place to place, if in fact they do, and recently had an opportunity to study this phenomenon in captivity. After decades spent observing the chimpanzees of the Yerkes National Primate Research Center, scientists suddenly saw the first instance of hand-clasp grooming. Although the Gombe branch-grasping technique had been seen many times, no one had ever seen a captive chimpanzee in any collection use hand-clasp grooming. After its initial appearance in 1992, the hand clasp began to spread. Initially the behavior was seen mainly in one female, but then other groomers began to adopt this tradition, and soon it permeated the Yerkes community. Five years later, even though the inventor of the technique was no longer living in the colony, the hand-clasp style had spread widely and its average duration had increased markedly.

Culture Slips into the Sea

The trouble started as soon as the word "culture" was associated with the ocean. The debate now includes biologists, philosophers, psychologists, and anthropologists. Two whale scientists, Hal Whitehead and Luke Rendell of Dalhousie University in Nova Scotia, are among those who defended the idea that cetaceans can possess cultural faculties, something extremely rare in the animal kingdom. Their review paper in the *Journal of Behavioral and Brain Sciences* ignited a storm of reactions regarding the existence of cultural processes in the sea and marked the beginning of a new era of cetacean culture studies.

As we have already seen, learning and imitation are important in defining our human concept of culture. A good example of social learning in the ocean world is the vertical cultural transmission from dolphin mothers to their offspring of foraging and feeding specializations. Besides the sponge-toting behavior of the Shark Bay bottlenose dolphins, there are many other instances of this type of social learning passed from one generation to the next. At Monkey Mia Beach on the Australian coast, "friendly" dolphins beg humans for food, a unique example of a foraging strategy transmitted by adult bottlenose females to their offspring over at least three generations. At this white sandy beach filled with holiday visitors, dolphins have adopted one of the laziest techniques on the planet to find food. They swim near shore waiting for one of the local rangers to hand them a fish. On several occasions, calves have been observed swimming next to their mothers and learning the practice of begging themselves. Unfortunately, this behavior is not devoid of consequences. Research studies on this begging technique adopted by dolphins show that calves born from provisioned females had twice the mortality rate of calves born from females that find food for themselves.

Another culturally transmitted foraging method observed in bottlenose dolphins is the coordinated teamwork of dolphins and fishermen at Laguna, off the coast of Brazil. Much like the case of the Imragen fishermen and dolphins of Mauritania, mentioned in Chapter 6, the human-

dolphin relationship has continued from generation to generation, dating back to 1847. At Laguna,
this joint effort is similarly carried out, but with
even more precise communication. Twenty-five
to thirty local dolphins drive fish toward the fishermen. As the fish move closer and closer, the dolphins perform a distinctive rolling and diving behavior as a sign for the fishermen to cast their nets
in the water. Because not all fish end up getting
caught, the escapees become easy prey for the dolphins. The interesting thing, from a cultural point
of view, is that this human-dolphin teamwork is
learned and adopted only by young dolphins
whose mothers participate in the coordinated interaction.

Bottlenose dolphins are not the only ones to
exhibit vertical cultural transmission; other cetacean species, including belugas and killer whales,
do so as well. Young belugas on their first migra-
tion follow their mothers between the breeding and
feeding grounds, then repeat the same route on
their own throughout their lives. Eventually, the
young females teach the maternal migratory tradition to the next generation. This type of cultural

transmission is also well documented in several species of baleen whales such as gray whales, which migrate along a ten-thousand-mile route from the cold Arctic foraging grounds to the warm-water calving lagoons of Mexico's Baja Peninsula.

Transient killer whales display a more theatrical example of cultural transmission in their premeditated attacks on pinnipeds resting on land. Adults may encourage their offspring to learn the hunting techniques by pushing them up and onto the beach toward the seals. Sometimes, after the prey is wrestled back into the sea, the adults continue to involve their calves in the handling of the seals, in most cases still alive, so that the calves may experience firsthand how to forage for food. In some instances, adult females have been observed rushing the beach with their offspring in complete absence of potential prey. From an outside perspective this may look quite useless and hazardous, but the mother always remains ready to bring the calf back into the water if trouble occurs. In the end, it's all part of the lesson.

In the same neighborhood, resident orcas are specialized fish-eaters that pass on their diverse prey-foraging strategies to their offspring. Resident and transient killer whales aren't as similar as in their cultural habits as we might assume. They don't interfere with each other's prey preferences or foraging strategies, and they have entirely different social structures and vocal emissions as well. The differences raise the possibility that whale culture could have affected their genetic evolution.

One of the finest arguments in support of cetacean culture comes from the study of resident killer whale vocal dialects. In their matrilineal world, each pod has a distinctive set of discrete calls that sound the same every time they are produced and seem to be stable over many generations. These sets of calls are known as dialects. Usually, resident pods produce between seven and seventeen diverse types of these discrete calls. Studying resident killer whales in the coastal waters of British Columbia, John Ford found that these dialects are passed down from one generation to the next through vocal learning and were observed to persist for at least six generations, strongly supporting the evidence of cultural transmission in this species.

Cultural traditions in killer whales off Vancouver Island may also take the form of a ritualized greeting ceremony reminiscent of an eighteenth-century royal court. When different resident pods of a community meet after a long separation, each pod lines up abreast at the surface facing the other group for at least ten seconds. Then, slowly, the individuals of each pod break their lines to approach each other. These cheerful celebrations can last for days with animals mingling, playing, and rubbing against each other.

Working with captive bottlenose dolphins in the 1960s, David and Melba Caldwell first observed and recorded dolphin "signature" whistles. A dolphin's distinctive signature whistle is learned, highly influenced by its environment, and maintained by an individual throughout its entire life. The signature whistle is much like a name that allows members within a school to recognize each other by sound. This whistle "contour" is individually distinctive and stereotyped in some of its acoustic features. Playback experiments recently conducted by another proponent of the signature whistle theory, Laela Sayigh, showed not only that dolphins were able to discriminate among different signature whistles, but that they used them to recognize individuals, which helped the group stay together. When a dolphin communicates its own signature name, another dolphin often repeats it. In a dolphin world, this may be a way of acknowledging the other's presence.

In Sarasota Bay, young dolphins appear to learn their signature whistles from others of the same species, showing that these sounds are not innate. Male calves produce signature whistles similar to their mothers', whereas female calves have more unique sounds. These more individualized signatures among females may be helpful for avoiding identity confusion, especially toward the end of the weaning period when the females of a group tend to stick together. The problem doesn't exist for young males, because they usually leave their mothers when they mature, so they are unlikely to be misidentified.

Recognizing each other by name was something believed to be unique to our human culture, like our ability to distinguish different rhythms and music. Recently, however, scientists at the Disney Epcot Center were able to train dolphins to sing a phrase from the theme of the original *Batman* TV series. Captive bottlenose dolphins in Florida were not only capable of distinguishing rhythms but also able to reproduce them in the correct order at the correct moment. How did they accomplish this seemingly ridiculous task? An object, in this case a Batman doll, symbolized a specific rhythm-vocalization combination for the dolphin; when the trainer showed the doll to the dolphin, the animal immediately responded with a sound. The result was a high-pitched version of what some of us remember from childhood: "Batmaaaaaaan." Maybe dolphins are not yet ready to reproduce a Mozart symphony, but it seems that at least some elements of human music such as rhythm, pitch, and timbre may be recognizable in their underwater world.

Even considering the widely studied vocalizations of sperm whales and humpbacks, which lie outside the scope of this book, the sum of information available on cetacean culture doesn't begin to reach the level of knowledge achieved for primates. What we have learned about these aquatic animals, however, is that culture in dolphins exists, at least in some forms, and that multiculturalism (as observed in sympatric pods of orcas displaying different cultures) and cultural stability over several generations are also present in these animals. Multiculturalism and cultural stability can both have an effect on genetic evolution. Dolphins that successfully learn or inherit "successful cultures," such as a highly efficient pinniped hunting technique, will probably pass on their "good genes" to the next generation, affording them a higher chance of survival in an ever-changing and challenging environment.

The Culture Club

Whether we are thinking about chimpanzees or dolphins, there are a few questions we can ask about the evolutionary importance of culture. Where exactly does culture come from—does it evolve? Why are some traits selected and perpetuated while others are not?

First, the geographic distribution of cultures is difficult to explain. Few chimpanzee tool cultures follow geographic lines. Termiting sticks in eastern Africa and stone hammers in western Africa are rare cases of cultures that are more or less restricted to regions. Other tool cultures, such as leaf sponges and ant-dipping, show up in a crazy quilt of occurrence all across the African continent. This could be due to a mosaic pattern of cultural innovation, in which one creative chimp in Tanzania invents a new tool use on her own, and another happens to have the same thought in Ivory Coast two thousand miles to the west. Or it could be due to a patchy pattern of cultural extinction. Suppose tool cultures arose independently all the time, and all across Africa, but had a tendency to disappear with the deaths of the inventors. This pattern of selective extinction would produce an identical pattern to the one we see. So do we assume that cultural diversity in chimpanzees is the result of innovation, extinction, or some of each? There is no way to attack this question currently.

We can, however, ask how and why certain cultural traditions, whether technological or social, arise and spread. Biological evolution occurs primarily via natural selection. It is enabled by the transfer of genetic material from one generation to the next. It is also an inefficient process, because of the generation time required for genes to pass to the next generation, and because each reproductive act requires (in all higher animals) the reshuffling of genes from mother and father. If a cultural trait, such as a new way of using a tool, confers on its user a better chance of survival and enhanced reproduction, then it has a good chance of being passed on. Even though the innovation is cultural, not genetic, the tradition of its use, once established, should spread, supplying an advantage of evolutionary fitness to the inventor. Thus an entirely nongenetic feature could have a long-term genetic, evolutionary effect on the survival of a species.

Cultural transmission is a more powerful means of changing the makeup of a species than biological evolution. It is faster acting and responds more quickly to changing environmental circumstances. Cultural "evolution" does not require the massive shuffling of the genetic deck that slows down the rate of change to a glacial pace. Only a few groups of animals on this planet exhibit these sorts of cultural traits. Higher primates are certainly cultural animals. Cetaceans also exhibit some elements of culture. In both cases, the capacity for culture is an outgrowth of the evolution of a big, sophisticated brain. So to bring together the disparate threads of social complexity, intelligence, and behavior that we see paralleled in great apes and dolphins, we must tackle the issue of the rise of intelligence head-on.

chapter eight

TOWARD THE ROOTS OF
HUMAN INTELLIGENCE

THE PARALLEL EXISTENCE of big brains and certain aspects of the behavior and ecology of chimpanzees and dolphins is by no means a random coincidence. The ancestral hominid stepped out of the trees and walked upright into the future. The ancestor of the dolphin walked on all fours and lived by the swampy edges of lakes. Taking dramatically different paths, these two lineages adopted parallel adaptations that have resulted in evolutionary success. They are a study in striking convergences and stunning contrasts. It's time for us to bring these disparate threads together to see what evolutionary factors led to the parallel suites of intelligence and social complexity in apes and dolphins. Although nothing in the evolution of life is inevitable, it is unsurprising that two such different big-brained creatures share so many traits.

Dexterous Hands

Consider the contrasts between the evolutionary history of apes and dolphins: how and why they have taken such divergent paths? The ancestors of modern apes and humans, some 20 million years ago, were ape-like creatures that bore little resemblance to their modern descendants. The earliest apes did not have rotating shoulders, walk on their knuckles, or possess particularly big brains. In fact, scientists only recognize these primitive apes by one anatomical feature that reveals their linkage to living forms: their molar teeth. The nascent apes possessed molars with the same Y-shaped pattern of fissures linking five cusps that modern apes and humans share, hence their nickname: the dental apes. These creatures proliferated and diversified between 20 and 10 million years ago. From one such ancient ape lineage came the direct descendants of modern apes and humans.

These apes were highly adapted to life spent partly in trees and evolved a rotating shoulder apparatus that allowed them to grasp branches and hang while feeding. This is the arm-hanging and swinging adaptation that quarterbacks and gymnasts still put to good use today. Because the ape's most ancient primate ancestors 40 million years earlier had a grasping hand, they had fine-tuned manual dexterity. The grasping hand was a holdover of a key adaptation seen in the most primitive primates; it may have evolved for the dual purposes of grasping branches while nimbly moving through trees, and for reaching out to snare small prey such as insects or frogs. It evolved in concert with another basic primate trait—stereoscopic vision. When leaping through trees, accurate depth perception prevents a nasty fall to the forest floor.

This called for grasping hands and the visual acuity to reach out and grip branches. It may also have helped early primates reach out to capture small insect prey. The same grasping hand was still around tens of millions of years later when it became available for tool-using hominids. This contingent history, so typical of the mosaic of evolved traits that comprise the human body today, belies the notion most people have that animal species, including humans, were put together by evolution as a package. Instead, pieces are tweaked every generation, always building upon the earlier version.

So the ancestral apes possessed the dexterous hands and fingers they had inherited from the most primitive primates. An ape's fingers are long, and the thumb and other fingers just barely meet. This is in stark contrast to a human's thumb and fingers, which can easily be made to meet at a grasping point. As early humans evolved, the relative lengths of the fingers and thumb changed so that manual dexterity was increased, presumably for the benefit of tool manufacture and use.

The result of this manual evolution was a set of grasping hands with adept fingers that became the platform, with the brain as the guiding light, for technology. This evolutionary stage is one through which dolphins, of course, did not pass. Grasping hands were never an aspect of their evolution, nor a need of their behavioral niche. The hands of the emerging hominids evolved as the brain became incrementally larger, which enabled more and more sophisticated and subtle uses of the hand. In a sense, Darwin got it right when he speculated in 1871 that our upright posture freed the hands for tool-making, which placed a new premium on technological know-how. Although he did not know that our upright posture evolved millions of years before tool use appeared, there is no question that tool-making had survival and reproductive advantages. Such advantages played a role in the further expansion in the size of our brains. By 2.5 million years ago, we have strong evidence in the fossil record that early humans in East Africa were using simple stone tools to crack open the bones of animals they had either hunted or scavenged.

Our ape heritage set the stage for the evolution in humans of a higher form of intelligence than the planet had ever seen. There was nothing pre-ordained about it; we simply inherited the right package of traits, and once this toolbox was together, natural selection produced intelligence as a startling and new mode of the adaptive process. The randomness of associated traits—grasping hands, big brain, bipedal posture—made it possible. The ancestor of the modern dolphin or whale had none of these, and that has made all the difference in the fates of our two ancestries.

In the twenty-first century, the age of miniaturization, our hands have become limiting factors once again. This time, it is because our large fingers cannot negotiate a keyboard as tiny as computer engineers can easily produce. So many aspects of a modern computer—first and foremost the keyboard on which I write—are designed with the relative lack of manual dexterity in mind. In terms of the evolution of our hands, our ancestors were ahead of the curve in the Pliocene, but we have fallen behind the curve of technology today.

Gloved Hands

Last winter, I lectured on the life history of marine mammals to a group of about fifty undergraduate students at the University of California, Los Angeles. Dolphins, more than any other creature, seemed most to draw the attention of my young crowd. One morning, I arrived in class with the skeleton of a bottlenose pectoral limb to use as a visual aid in explaining the evolution of these ocean dwellers from their terrestrial ancestors. When I placed the skeleton on my desk for examination, my students seemed utterly surprised that the familiar hydrodynamic gray flipper that dolphins use to steer as they glide through the ocean was now presented to them in the form of an elongated, curved bony hand, not so dissimilar from their own. The skeletal flipper served as a vivid reminder that, despite their watery environment, dolphins are mammals through and through. On more careful examination, the differences between a dolphin's "hand" and our own became apparent. Dolphins have more phalanges in this limblike structure as an adaptation for the specialized "fin" functions that steering through water requires.

At roughly the same time that the earliest primates were emerging from other mammalian stock and the grasping hand first appeared, ances-

tral dolphins began their big break, too. The cetaceans evolved from hoofed land mammals, which
gave rise to the archaeocetes, an early, doomed
branching of precetaceans. More than 50 million
years ago, the ancestors of dolphins were true land
animals about the size of a dog. In their transition to an aquatic existence, the limbs of these ancient dolphins evolved into a miniature version of
their terrestrial legs, retaining the same number
and arrangement of hind limb bones as their ancestors had.

But how do we know that modern dolphins
come from furry, four-legged mammals? Hans
Thewissen suggested that one way to understand
the evolution of these animals is by their embryonic development. Over 15 million years ago accu-
mulated genetic changes accelerated the pace at
which cetacean limbs shrunk, generation by generation. The absence of one gene in particular was
identified as responsible for the decline and loss of
the hind limb: the sonic hedgehog gene. This powerful gene is known to facilitate normal limb development. About 15 million years ago, the sonic
hedgehog gene disappeared, leaving dolphin em-

bryos to develop hind limb buds instead of legs. Thewissen explains that "the genes are similar to the runners in a complex relay race, where a new runner cannot start without receiving a sign from a previous runner. In dolphins, however, at least one of the genes drops out early in the race, disrupting the genes that were about to follow it." Thewissen's developmental data from embryonic spotted dolphins, together with other fossil and DNA information, offer us insight into the evolution of the perfectly hydrodynamic animals that we observe swimming in the oceans of today.

One of the most obvious differences in gross anatomy between dolphins and apes is the absence of hands or feet in dolphins. This adaptation has certainly limited the ability of cetaceans to evolve technological skills—inventing something with a big brain is futile unless there is a way to put it together. Throughout the history of early humans, the ability to make and use ever more advanced tools increased and at some point played a key role in the expansion of the brain. The other constraint on the evolution of primate-like intelligence in cetaceans is the medium in which they live. Water exacts a tremendous cost just in maintaining body temperature. Its vast depths and murky translucency also present problems for visual communication. Dolphins used to be thought of as visually impaired, relying primarily on echolocation; this has in recent years been shown to be simplistic. Several species can see quite well underwater and even out of the water. But the lack of hands for grasping is their eternal drawback in the evolution of primate-like intelligence.

Brain Size and Intelligence

Shortly before the archaeocetes became extinct, the precursors of modern whales and dolphins emerged. About the time that apes were making their first appearance on Earth, the earliest recognizable dolphins arrived too. And by the time hominids evolved from their ape progenitors, modern species of whales and dolphins had evolved. But whereas the primate brain was built by natural selection in several obvious lobes, each with a specialized set of functions (with much overlap), the emerging dolphin brain bore little resemblance to this model. It is instead composed of several tiers of tissue comprising some of the same functional regions we see in the primate brain and others whose functions are still unclear.

Dolphin brains are highly convoluted and larger in both volume and mass than human brains. The average adult human brain weighs about 1.3 kg, whereas a bottlenose dolphin has a cerebral mass that is about 25 percent heavier. So if intelligence were measured by brain size alone, sperm whales would be the smartest animals on the planet, with a brain weight of 7.8 kg, followed by African elephants, weighing in at 7.5 kg, then fin whales at 6.9 kg. Humans would clock in just above cows. Because large brains are often related to large bodies, a biologist at the University of California at Los Angeles, Harry Jerison, devised the encephalization quotient (EQ) index, which expresses brain size in relation to body size. In this metric measurement, humans regain our familiar spot at the top of the thinking short list of Animalia. One interesting result is that, comparing the EQs of different anthropoid primate and dolphin families, similar levels of encephalization emerge in extremely diverse phylogenetic lineages. Further, both dolphins and apes have very large brains in relation to their total body size, placing them just behind us on the EQ list.

Here's an even more astonishing fact: the level of encephalization of *Homo erectus,* the direct precursor of *Homo sapiens,* was not much greater than that of the tucuxi, a coastal river dolphin found today in the murky waters of the Amazon Basin. Looking further back in time—2 million years—the most highly encephalized mammal on the planet wasn't a hominid, it was a dolphin.

Dolphins have had bigger brains for a longer time than our species has, so why don't they dominate Earth? What set us apart from dolphins was the tremendously rapid enlargement in our brain size over a relatively short time period. Neuroanatomical studies tell us that dolphin and human brains share similarities in the presence of asymmetry and behavioral "laterality," known in humans for being associated with such complex cognitive processes as language. Recent research shows that the cetacean cortex, previously believed to be quite uniform, is instead rather diverse, creating a series of implications for cognitive complexity. Despite sharing the attributes of complexity and diversity, however, there are significant morphological differences between dolphin and human brains. Just to cite a few, the overall brain shapes are different, the cortex thickness is much thinner in dolphins, and there is a greater proliferation of tissue in the temporal-parietal areas of a dolphin and a higher degree of cortical folding.

These and other differences between the brains of dolphins, great apes, and humans are probably evolutionary responses to different environmental demands. But in both great apes and dolphins, the large brain confers on its owner the ability to parse out complex social dynamics in the group—to be political animals, literally.

The Ties that Bind

Now that we have examined the key differences in our evolutionary pathways, what are the points of convergence? Why did brainy intelligence take hold in these two lineages that seem to share so little else?

Some species of apes and dolphins share many behavioral similarities: a tendency to live in fluid and flexible social groups that may split up and come together unpredictably throughout the day; that famously big brain; a diet that tends to be composed of widely scattered, high-quality foods—fruit or fish; and a tendency for males to form tight coalitions that they use to coerce other males and sexually "persuade" females. These fascinating convergences make it clear that natural selection, working in a finite world, produced a finite number of potential forms. We already know well that environments that are similar in obvious ways tend to have a similar range of ecological roles, or niches, available, and so animal species only distantly related may look strikingly similar in similar environments. Mammals in Australia look very much like mammals in the rest of the world, until you look closely and see that the Australian forms—marsupials—have reproductive systems that have been engineered in an entirely different way—with a pouch instead of a placenta, and a penis that passes only sperm while a primitive cloaca handles urination. Superficially, there is stunning convergence: marsupial wolves (the enigmatic, recently extinct thylacine) and placental wolves; marsupial mice and placental mice. In the fossil record there are marsupial lions. Even the modern kangaroo is similar to a placental deer or antelope (both of which are nonexistent in Australia). Although kangaroo and antelope evolved to have superficially different body plans, both exploit the same resource base—grasslands.

Encephalization, the expansion of the neocortex of the brain through the course of evolution, is a key adaptation of both chimpanzees and dolphins. It is part and parcel of the way in which they respond to environmental pressures. And although the underwater world of the dolphin and the rain forest habitat of the chimpanzee may not seem very much alike, the environmental pressures both species face and the tactics they use to respond to those pressures are surprisingly similar. Chimpanzees learn the locations of hundreds or thousands of fruit trees, remembering when they are fruiting and when they are not. Their range is vastly larger than that of lower animals that eat the same foods but forage with a more hard-wired

plan. What natural selection has favored in chimpanzees is flexibility—the capacity to modify their
behavior on a moment's notice in response to
changing food supplies. They use not only their
sense of smell or hearing to find food, they also use
their big brains to fashion novel ways to extract
food from their environment through the use of
tools.

Dolphins also rely on their brains and their behavioral flexibility to find food. Species like orcas,
bottlenose dolphins, and common dolphins are
opportunistic, switching food sources and adapt-
ing to new circumstances depending on prey availability.

In the coastal waters of the Amvrakikos Gulf
and around the island of Kalamos in western Greece,
bottlenose dolphins have developed diverse tactics
to cope with very different environments within a
geographically small area. In the semi-closed waters of the gulf, which are quite polluted from river
runoff and other contaminant sources, bottlenose
dolphins usually display surface foraging and cooperative feeding. But in the open and less polluted waters around Kalamos, the dolphins con-

centrate their feeding on fish living at greater depths, with little or no surface feeding.

Wherever you encounter bottlenose dolphins around the world, you'll find an opportunistic species able to adapt to a variety of circumstances. For example, the behavioral plasticity of these animals allows them to capitalize on human activities, like the cooperative coastal fish netting of humans and dolphins in Laguna and Mauritania, described in earlier chapters. Such solutions require the flexible, learned behavior repertoire that only intelligence can enable.

This sort of cognitive adaptation is certainly not unique to either apes or dolphins, but it finds one of its highest forms in them. However, large brains require a slow life history. Humans spend well over a decade in the process of growing up and learning how to become human, the process we call socialization. Chimpanzees also spend more than a decade maturing. During this time they are constantly uploading behavioral "software," as it were, to help them respond to all the nuanced and novel situations that will confront them in life. This is the tradeoff; lower animals mature faster and don't need all the learning that can come only with experience at living. Humans and a few higher animals incorporate a high percentage of learned information into their survival and reproductive strategies, which requires years of maturation. Recent work by a team headed by Luis Lefevbre examined the relationship between brain size and different periods of life history in males and females of twenty-five species of odontocetes. The results showed that, in a way similar to their distant primate cousins, brain size has co-evolved as life history periods extended, meaning that a slow life history is a significant component of encephalization.

But why is a big, sophisticated brain an advantage in life? Dinosaurs had puny brains but flourished for hundreds of millions of years. Intelligence is an evolutionary adaptation, but not necessarily the only or even the most effective one. What works depends on the environmental context. Some creatures have been on Earth for hundreds of millions of years and have changed little. Other lineages, like primates and cetaceans, have undergone dramatic changes and a mushrooming of brain size in just a few million years.

Natural selection acts to favor intelligence when intelligence confers survival and reproductive benefits that compare favorably with traits that are genetically hard-wired. The ability to make and use a tool is an example of such an adaptation among chimpanzees. The tool is not the adaptation; the ability to learn and imitate is. The ability to respond to rapidly changing dynamics in the social group, such as by males forming coalitions to control females, is not limited to higher primates and dolphins but certainly typifies many species among them. In each case, these skills require years of learning, but the payoff is a potential reproductive windfall. Tools allow chimpanzees to harvest protein, fat, and carbohydrates otherwise unavailable. The added nutrition can help to get a female who is gestating or lactating through an otherwise lean time of year and enhance her reproductive output over the course of her long life.

The Shakespearean dramas that unfold in both chimpanzee and dolphin societies when males attempt to manipulate females rely on the tactical cleverness that only a big brain can muster. We used to think that such male alliances were based entirely on kinship. Now we know that males form such coalitions for mutual benefit, only to turn upon one another after overthrowing an alpha. Bonding through cerebral sophistication is a commonplace tactic for survival among humans, and it is just as important for chimpanzees. Male dolphins likewise employ such coalitions to drive off unwanted male competitors and to secure sexual access to a female.

The large brain may also help in foraging. As we've seen, both chimpanzees and dolphins feed on widely scattered, temporarily available foods. Dolphins chase schools of fish, chimpanzees chase the fleeting appearance of ripe fruits in tree crowns. These two dietary specialties keep both creatures moving all day long in search of the next school, the next patch. But predicting where and when to search is the hard part. Chimpanzees have the spatial memory of a forest ranger, monitoring particular fruit trees in the weeks leading up to the ripening of a crop, then returning to the right spot day after day until the bounty is gone. Dolphins have a taller order; they have to know where to locate rapidly moving fish schools and to do it without such obvious landmarks as trees, streams, and mountains. For this they have sonar, a wonderfully evolved system that humans only recently were able to replicate for their own uses. But sonar alone will not provide a meal; dolphins put their own intelligence (and memory) to good use in finding fish and catching them.

In different areas of the world, bottlenose dolphins utilize entirely diverse strategies that include hunting in a team, foraging individually in the surf zone, pinwheeling (encircling the prey to cut off its escape), fish-whacking or fish-kicking (encircling prey, then lashing their tails through the middle of the circle, launching stunned fish into the air), chasing fish schools onto mud banks, following fishing boats, and cooperating with fishermen by herding fish into nets. How some dolphins put their own brains to work in "extracting" food from their environment is quite well studied, but understanding a wild dolphin's spatial memory is another story. The majority of "memory" studies come from animals in captivity. Trainers, sitting or standing at poolside, perform all sorts of tests and wait for the dolphins' responses. From experiments like these, it appears that spatial memory may play an important ecological role, but the deep open oceans still hold the secrets of how dolphins use their minds in their natural environment.

The parallel adaptive suite of traits shared between chimpanzees and dolphins does not prove conclusively that natural selection, working with a similar set of ecological and social pressures, produces similar cognitive means of coping with them. The dramatically different organization of the dolphin and human brain tell us that different evolutionary pathways can produce similar results. It certainly suggests that brain complexity, social complexity, and ecological complexity are linked.

Is Intelligence Universal?

Understanding the parallels between chimpanzees and dolphins is one thing. Extending the lessons of their similarities to the rest of the animal world is quite another. Do highly intelligent species on Earth share a set of universal characteristics? We can turn to the examples of intelligent life that we know well for clues. In addition to primates and cetaceans, we could point to elephants. These amazing creatures possess enormous brains, even in proportion to their outsized bodies. They have an appendage almost as dexterous as any hand, their wonderful trunk. And they have been shown to have problem-solving capacities beyond those of most other mammals. Many people will make a valiant argument for their pet dog or cat, citing anecdotes in which the pet seems to show it was almost human in intellect. Certainly the cognitive line between ourselves and other animals has grown fuzzier with increased research. But for the most part, we anthropomorphize when we attribute humanlike smarts to any creatures other than dolphins and great apes.

How many pathways might there be that lead to this level of intelligence? Dolphins and primates represent the tips of the many thousands of lineages of life on Earth, only a handful of which have produced highly intelligent creatures. But another way to look at the issue is that we know only one planet well—our own—and on that one planet we find humans plus a few other quite intelligent life forms. Taking this second view, we could argue that given the right set of physical environmental circumstances—a comfortable temperature range, water, oxygen, nitrogen—the evolution of intelligent life might be a likely eventuality rather than a bizarre fluke. Just as Australia presents us with a test case of natural selection, the constraints of the physical environment strongly suggest that only the right combination of environment, evolutionary baggage, and perfect happenstance timing produced humans. If the clock of Earth's history were run again, there is no reason to assume that humanlike intelligence would arise anew.

BEAUTIFUL MINDS ARE A TERRIBLE THING TO WASTE

IT HAD BEEN IN THE OFFING for quite some time. Incessant noise and chemical pollution contributed to a highly degraded habitat. There was illegal overfishing, both with explosive and toxic substances, and the unregulated and choking traffic of commercial vessels. Next came warnings from the scientific community spurring increasing concern from international wildlife organizations and the general public. But the alarm was sounded too late; in the Yangtze River of China, one of the most polluted waterways of the world, the last baiji dolphin was gasping its final breath.

The baiji, an endemic species that diverged from other dolphins over 20 million years ago, was considered by many to be a "living fossil." Its demise marks something we have not seen in modern history, the first recorded cetacean extinction caused by human activities. We have now witnessed what we have long talked about in conservation circles but never really expected to see in our lifetimes.

In the 1980s, around 400 baiji were living in the Yangtze. By 1997, a research expedition found only thirteen of them still living. In 2006, after six weeks of high-tech surveys along 3,500 kilometers of the Yangtze, a team of researchers came up empty-handed: there was no trace of the pale and nearly blind dolphins. At least twenty to twenty-five individuals are needed for this delicate species to survive, but not one has been found. So the baiji—locally known as the "goddess of the Yangtze"—is gone forever, a victim of our economic well-being, our ruthless disregard for our environment, and our lack of respect for other forms of life that share Earth with us. Now another species, the Yangtze finless porpoise, is on a similar deadly path. If we sit still and do nothing, the finless porpoise will likely be the second cetacean species in our lifetimes to disappear forever. The warning for humanity is sobering, and the loss is irreparable.

Glamour Species

The future of great apes and dolphins depends on the health of tropical forests and of the seas. As we reach farther and farther into tropical forest in search of timber, farmland, and village clearings, we remove habitat for countless living things as well as the animals themselves. As we continue to use the oceans as our dumping ground, we make the inshore areas in which many dolphin species live uninhabitable.

Do we want to save our cerebral cousins? Yes, of course we do. But conservation issues are never that simple. They are about costs and benefits—the feasible and the marginally unfeasible. Which species are saved and which are left to languish usually comes down to whether the species is high profile and whether saving it is economically viable, or even convenient. When a highly endangered, small species of moth or beetle lives in the path of development, it nearly always loses to human interests. The economic forces are simply too great for conservationists to make a strong case for saving a beetle when millions of other beetle species remain.

Fortunately, great apes and cetaceans are high-profile animals. They are distinctive, representing a tiny minority of the world's creatures that are extremely intelligent. Apes are also our closest relatives. Both dolphins and apes have captured our imaginations and hearts for hundreds of years. When appeals are made for conservation awareness and funds, the donating public is most likely to open their pocketbooks for apes and dolphins. The "smile" of the dolphin certainly does not hurt its appeal, either. They are, in conservationists' terms, "glamour" species. They are also near the top of their respective ecosystems. Dolphins are predators and among the larger creatures in their environments. Great apes are the largest of the non-human primates, and usually among the larger mammals in their forest habitats. This means that when you preserve a population of gorillas or of dolphins, you are incidentally protecting their habitats, which tend to be full of much less glamorous but often equally endangered species.

Protecting even the most glamorous of species, however, is fraught with difficulty.

In the Land of Conflicts

All great apes live in deeply impoverished, developing nations. They require large areas of tropical forest that are also desired by a rapidly growing human population and by the commercial interests of loggers and miners. Much of this tropical forest is not readily visible to the outside world. Only when outsiders make inroads into the Congo Basin can we discover the fate of the apes therein. Many critical great-ape habitats are in near-constant threat of civil war, which renders them inaccessible to conservationists for months or years on end. In Kahuzi-Biega National Park, in the eastern rain forests of the Democratic Republic of Congo, scientists waited while the Congo civil war raged for years. When they could finally venture back in to recensus endangered populations of the eastern lowland gorilla, they found a precipitous drop in the gorilla populations. Civil war means no protection by forest rangers, and it means many hungry, heavily-armed soldiers who can prey on the gorillas at will. War creates waves of refugees fleeing the fighting, and as they pass through gorilla habitat they kill and eat whatever they can.

Even a country with a proactive government can have difficulty protecting its endangered species. The double-edged sword of ecotourism can be seen most vividly in my own gorilla research site in Uganda. For many years, the pioneering gorilla researcher Dian Fossey ran her mountain gorilla research project like a fiefdom, threatening local African poachers at gunpoint and keeping foreign tourists out as well. After Fossey's death in 1984, the Rwandan government developed an ecotourism project that quickly became a major source of revenue for the impoverished country. Other populations of mountain gorillas, including those in my own field site in Uganda, also became tourist attractions and cash cows for the local governments. The initial result was effective protection of magnificent creatures that would otherwise have faced almost certain extinction.

Over the ensuing two decades, tour companies flocked to the realm of the gorilla. They built lodges and hired local laborers. The government trained guides, rangers, and guards to lead tours and protect tourists. Small towns popped up around the forest reserves where the gorillas lived. In the 1980s and 1990s, despite repeated coups and civil wars in Rwanda and Congo, there were always some mountain gorilla tour sites open, and everyone, including the apes, benefited.

Then came a much-publicized assault on Ugandan ecotourism by a local rebel militia. Tourists were murdered, others were terrorized, and the tourism project was shut down. Tourism in Uganda dropped precipitously because the country was perceived as unsafe for travelers. The tourism-conservation balance that had seemed so effective was in fact quite fragile. Two years later, tourists began coming back, and the military had beefed up its presence in the reserve to discourage future attacks. But the stigma of Uganda as a place where only the bravest should venture will linger for years, and the loss of tourism will hurt gorilla conservation efforts as a result.

Bush Meat

The great apes rely on large areas of undisturbed, healthy tropical forests for their survival. These forests are in rapid decline, and with them go the apes. In Indonesia, the rapid and often illegal cutting of rain forests has led to such a precipitous crash in orangutan populations that an estimated 80 percent of the world's population of Sumatran orangutans have disappeared in the past decade. In Africa, the loss of forest is only one problem. The hunting of gorillas, chimpanzees, and bonobos continues unabated in many parts of central and western Africa. The bush-meat trade is the commercial use of the meat of many kinds of mammals, easily seen in any local African market. But whereas antelope, pigs, and other animals breed quickly enough to sustain at least some bush-meat trade, apes reproduce so slowly that even modest levels of hunting can wipe out entire populations in short order. In an open air market in the Central African Republic or Cameroon, the twisted, blackened hands of smoked gorilla meat sit in baskets awaiting sale. The sales are brisk, and they are not to poor villagers. The local consumption of bush meat by villagers for thousands of years to add some protein to an otherwise bleak diet poses less of a problem than purchases by relatively well-to-do Africans who consider bush meat a delicacy and are willing and able to spend up to five times the cost of beef to obtain it. This demand drives the commerce, and recent advances in logging in Africa enable it. In the days when the African wilderness began at the end of the last road into the rain forest, apes had a refuge out of reach of all but the local villager's gun.

Today, international logging companies have cleared major roads into the deep interior of the rain forest for the purpose of extracting timber. These roads become thoroughfares for enormous trucks carrying logs out of the forest. They are also pipelines for the easy and quick transportation of bush meat to markets in larger towns and cities. Instead of supplying their logging crews with food, some European logging companies simply supply them with high-caliber guns and ammunition and allow them to shoot whatever they can. This has created a terrible irony. For decades the rallying cry of "Save the Rain Forests" has led us to believe that if the forests can be preserved, the wildlife will also be saved. But as a result of the logging roads, there are now entire expanses of rain forest that are nearly pristine, except for the complete loss of their large mammals. The desire for the meat of great apes is driving them into extinction even in areas where the usual threat of habitat loss is not yet a problem. The demand for bush meat as a high-status delicacy among Africans is a deeply held cultural value that has proved difficult to change. This demand has driven a black market for bush meat that extends to Europe and North America. If you know the right people, you can find smoked chimp and gorilla meat secretly for sale in Brussels, Paris, and New York.

As the vast rain forests of central Africa are

opened up, we find other troubling signs that great ape populations are in terrible trouble. Many of the same viruses that have emerged to threaten human populations are also hitting apes, with devastating effects. Ebola is suspected of killing thousands of gorillas and chimpanzees in recent years. A recent report revealed that anthrax occurs in some chimpanzee populations and may be responsible for the loss of entire chimpanzee communities.

Bloody Oceans

Dolphin populations are much less visible than those of great apes, but the threats to them are no less insidious. There are cetaceans living in the oceans about which we still know virtually nothing; several species of beaked whales—large dolphin relatives—are known only by the few individuals that have washed up dead on coastlines. We know more about the large species, the highly social species, and the ones that occur close to shore. Nearly every one of them, the dolphins we know and the ones that we still don't know, are either endangered, threatened, or of unknown status. Even if it is increasingly difficult to track the reasons why some of these species are so vulnerable, one common denominator is brutally clear: our human impact.

Hunting of Cetaceans

Commercial whaling is responsible for the disappearance and decline of a colossal number of marine mammals, especially large whales. North Atlantic right whales, so named by whalers because slow speed, large size, and a tendency to stay close to the surface made them the "right" whales to hunt, now rest at the brink of extinction, numbering less than 300 individuals in the wild. Their population has been so decimated that even though they are no longer hunted, ship collisions alone may be enough to wipe right whales from Earth forever.

Attempts to manage whale stocks and raise public awareness on the inhumane aspects and unsustainability of whaling range from the regulations of the International Whaling Commission, going back to 1946, numerous conservation organization campaigns, and the bold activist attempts of Greenpeace and others like them to interfere with whaling activities. Still, Japanese, Norwegian, and Icelandic whaling fleets perpetuate this blood business by continuing to hunt whales throughout the world. Recent statistics from the Wildlife Conservation Society show that 20,000 small whales and dolphins are killed each year in Japan alone. Intentional killing of small cetaceans is common in developing countries like Sri Lanka, Peru, and the Philippines, where dolphins are slaughtered and their meat is sold in local markets or used as crab bait. Keeping track of this butchery is next to impossible, let alone finding an effective way to stop it. The perpetrators tout their respective reasons for doing what they do (scientific studies or preservation of aboriginal traditions, for example), but the final result doesn't change. Because of hunting, the world's marine mammal stocks continue to face drastic declines.

Seafood Consumption

Recent figures indicate that nearly 80 percent of the world's fisheries are over-exploited or significantly depleted. This is one of the gravest prob-

lems facing the marine environment today. In preceding generations little attention was paid to
the destruction wrought to marine ecosystems. We
thought that the oceans, vast as they are, were immune to human impact and that fishery stocks
were essentially infinite. We now know that this
is not the case and that unregulated commercial
fishing has depleted fish populations to a degree
that some of them may never recover, with an ad-
verse impact of unknown magnitude to the ecosystem overall. Further, cost-effective commercial fishing methods, especially deep-sea bottom trawling,
convert thousands of square miles of productive
ocean floor into a desert-like wasteland, incapable
of supporting life.

One shocking example of overfishing is the collapse of cod stocks off Nova Scotia. Cod were once
so plentiful in this area as to be considered an
inexhaustible resource, providing generation after
generation of professional anglers with a stable, re-
liable livelihood. Since the 1850s, the population
of this top predator has plunged 96 percent according to historical fishing records and archeo-
logical data. The decline of cod over the last few

decades has changed the entire local marine eco-system so much that, according to many scientists, it will be extremely difficult for this species to re-cover, perhaps impossible. The cod fishery is a case in which nonsustainable fishing practices have caused deep, everlasting changes in the ocean envi-ronment with potential repercussions likely to cas-cade all the way down the food chain. Such reper-cussions can be seen not only in the underwater environment, but also in human communities that have relied on commercial fishing for hundreds of years.

One example of overexploitation and its ef-fect on cetaceans comes from the Ionian island of Kalamos, where my brother Giovanni Bearzi—also a field cetologist—and his team from the Tethys Research Institute have actively studied common dolphins since 1996. Back then, tuna were so abun-dant in the area that, from a distance, they were easily confused with the plentiful dolphin groups of the time, feeding on schools of anchovies and sardines near the surface. Those days are now gone, and common dolphins, tuna, and swordfish, as well as anchovies and sardines, have become rare in the area. In a shockingly short period, fishing with unregulated purse seine netting has virtually wiped out the epipelagic fish stocks in the area, which in turn has impacted the larger species that depend on them for sustenance. The dolphins' search for prey has become increasingly difficult, causing them to disperse over a wider area and forcing them to become more nomadic. They have broken up into smaller groups, and their encounter rates have decreased twenty-fivefold in only eight years. Today, Mediterranean common dolphins, once considered the most regular cetaceans in the region, have found their place in the growing list of endangered populations.

What we know is that overfishing has placed us—and dolphins—in a peculiar predicament in which we are now fishing our way down the marine food chain. We harvest smaller and smaller fish that do not live long enough to reach reproductive maturity (thereby restoring their numbers). We now eat species that we used to discard or ignore. The question that emerges from our behavior is how far down the chain we can go before the whole system fails. Respected oceanographers and marine biologists like Jeremy Jackson, Daniel Pauly, Elliott Norse, and a host of others would argue that the ecosystem has already failed and that in many areas, the damage is irreversible.

Careless and Inadvertent Killing

Bycatch—the incidental catching of nontarget species in commercial fishing activity—is responsible for the deaths of millions of dolphins in recent years alone. Where overfishing threatens the ecosystem overall, bycatch is one of the most severe problems facing dolphins worldwide. Drift netting, a commercial fishing practice that utilizes long nets set and left to drift with the currents, target anything that happens to swim into them. This indiscriminant method kills countless marine mammals every year. For example, a recent case study led by Sergi Tedula looked at 369 fishing operations targeting swordfish in the Alboran Sea. In a one-year period these operations harvested just under 3,000 target swordfish and in the process killed over 235 dolphins. Estimates for the entire driftnet fleet in the same twelve-month period projected between 3,110 and 4,184 dolphin deaths in the Alboran Sea alone, and between 11,589 and 15,127 dolphins killed just around the Straits of Gibraltar.

Another dramatic illustration of bycatch is the effect of gill netting and shrimp trawling on the vaquita, a small harbor porpoise found in the Gulf of California and one of the cetacean species closest to extinction in the world today. The decline of this species has been traced directly to unsustainable fishing practices. With only 224 animals left on the planet, this porpoise is now listed as an endangered species in both the United States and Mexico.

Dolphin bycatch is also well documented in the Eastern Tropical Pacific, where yellowfin tuna are known to follow dolphins in order to find food. Because of these associations, fishermen intentionally circle dolphins in order to catch the school of tuna swimming just beneath them, and dolphins often end up in purse seine nets. This technique, called "fishing on porpoise," has been used since the 1950s and is responsible for catastrophic declines in some dolphin populations.

Pollution

Pesticides, insecticides, and toxic elements flow from the hearts of our cities down to the sea. Then, hand-in-hand with the currents, they travel to the most unexplored waters and the deepest ocean canyons. For an ocean dweller, there is no escaping pollutants. Dolphins are among the animals most affected by manmade contaminants. Simple organisms absorb small quantities of a pollutant, but when simple organisms are eaten by bigger organisms, the pollutant levels increase up the food chain in a process called biomagnification. The highest concentrations of pollutants, therefore, are found in large, long-lived predators like dolphins that feed at the very top of the chain. Contaminants are transferred through the rich milk of the mother directly to her offspring, bequeathing a series of dire consequences. Although the effects of these substances are hard to quantify, there is growing evidence that calves may accumulate high concentrations of PCBs, DDTs, and other pollutants that can cause reproductive deficiencies in adulthood or even death.

Marine debris raises yet another problem. About 100,000 marine mammals die every year by suffocation or drowning from entanglement in plastic debris. Spending days at sea as a marine biologist, I sometimes feel like a street-sweeper in a big city, especially after large rainfalls. We come across all kinds of nonbiodegradable trash in our ocean research, sometimes in a seemingly endless stream that may last for days. Marine debris is found everywhere in the ocean. There are about 46,000 pieces of plastic debris alone floating at or near the surface for every square mile of ocean. This material tends to accumulate along fronts (interfaces between diverse water masses), where seabirds and dolphins often convene to forage. The price of feeding in these areas can be deadly. Practical solutions to our dependence on plastics may be too far in the future to help dolphins.

Greenhouse gases raise another set of problems for dolphins. Climate changes can occur with or without human help, but in recent years, we have drastically accelerated the production of

greenhouse gases, warming the planet to dangerous levels. The effects are already measurably real.
Shrinking of polar icecaps and increases in ocean
temperatures can interfere with the habitat of cetaceans living in many parts of the globe, because
shifting currents off the U.S. West Coast and in
the Barents Sea in the Russian Arctic have affected
fish populations. Some species have increased while
others have decreased drastically. This has negative
ripple effects from cetacean diets on down the entire marine food chain.

If greenhouse gases, chemical pollutants, and
debris aren't enough, we can also claim responsibility for making the oceans a lot noisier. Dolphins
and whales live in an acoustic world and rely on
communication for their well-being. Loud sounds
from Navy sonar experiments, oil exploration, underground blasting, and ship traffic can harm these
animals in ways often difficult to quantify.

Unintentional Harassment

I recently took a group of students to observe a
large school of spinner dolphins off Oahu, Hawaii,

for a field class. We weren't alone with the dolphins for very long before three gigantic yellow inflatable rafts full of whale-watching tourists arrived at our site. One after another, the whale-watchers plunged into the water to swim with the dolphins. Within a few minutes, the previously tranquil bay was transformed into a roiling mass of multicolored humans, harassing the dolphins, swimming upside-down to inspect and touch their smooth skin. So ended the quiet afternoon for the spinners. The dolphins were hit with swim-fins and forced into compact groups, moving from one spot to the next in an attempt to avoid the human onslaught.

Swimming with dolphins is big business, to the tune of millions of dollars spent worldwide every year. From the human perspective it would seem an incredible experience, perhaps bordering on spiritual, if one is prone to that way of thinking. But let's look at what it might be like from the dolphins' point of view.

Dolphins are wild animals and have needs that must be fulfilled for their survival and well-being. Resting plays an important part in their life rou-

tine, which was precisely what they were doing before the whale-watching well-wishers showed up.
To the dolphins, the tourists were a disturbance, as
evidenced by their attempts to swim to less crowded
parts of the bay. If the disturbances continue, as
they surely will, the animals will need to find a new
area to rest. The consequences of this are difficult
to quantify, but such harassment is likely not beneficial to animals. The Oahu dolphins inhabit U.S.
territorial waters, where the Marine Mammal Protection Act is in full force, theoretically protecting them from this type of exploitation. Why then
does this occur?

Do dolphins really want to swim with us? Their
eternal smile may seem friendly, but their behavior often tells us that the best way to commune
with these animals may be to observe them from a
distance.

Filippo is dead. On August 6, 2004, just after I finished the first chapter of this book, his body was
found a mile outside the old port of Manfredonia

in Southern Italy. Nicola Zizzo, in charge of the autopsy, found that the dolphin had died from trauma caused by the fractures of several right and left ribs that pierced and wounded the internal organs, causing heart failure. From Filippo's dorsal region, the doctor also extracted several small-caliber bullets, likely to have been in the body for at least a few years.

By the day after this sociable dolphin was found, he was completely dissected, sealed in sterile vials, and sent to the lab for further analyses. Once the analyses were complete, Filippo's remains would be thrown in the trash. The city of Foggia launched an investigation into the facts surrounding the death of this now-famous animal and found that the likely culprit was our own species, most likely a bomb thrown by a fisherman.

The story and life of Filippo, the perhaps too-human dolphin of Manfredonia that used to play and follow fishing boats on their way out to sea, ends here; between those who loved and respected him, those who harassed him, and those who didn't do enough to keep him alive.

Filippo's story is one of many about "orphaned" dolphins that, for one reason or another, have chosen to live near humans. Their proximity to our species has made some of them famous the world over. There was Jean-Louis, which lived off Cap Finestere in Brittany; Beaky, an old male which lived near the Isle of Man in western England; and Simo, a juvenile that turned up in Wales off Pembrokeshire. There were Percy, Dolphy, Fanny, Dolly, Marine, Filippo, and many others. Each of these dolphins had its own story, its own unique personality and ways of interacting with us. But what these animals had in common was that they were victims of our increasing presence and our incessant desire to have contact with them. Sadly, one thing that we still don't seem to be able to grasp is that all of these animals need their space, both physical and emotional, just as much as we need ours.

Best in Show

Today, there are hundreds of Flipper-like dolphins kept in captivity worldwide, many of which die before their time in these facilities.

The ethical question of keeping dolphins in captivity is a hard one. There are several arguments in support of keeping them in captivity and several against it. Captive facilities provide opportunities for close observation by masses of people. They educate and raise public awareness, and they offer an excellent platform for conducting research and provide information useful in conservation efforts. There are some who believe that high quality oceanaria offer these animals a better life than the one they might spend in the open ocean.

Those against captivity believe that the removal of dolphins from their natural environment means more dolphins taken from the wild, and point out that they don't survive as well in captivity as they do in nature. They believe that the chlorinated tanks of captivity represent nothing more than a jail for the animals, stifling them acoustically and spatially. The anti-captivity folk also bring a moral question to the table, believing that dolphins have rights equivalent or close to our own, and to keep them in captivity is an abuse of those rights, regardless of how well they might be treated.

Dolphins are close to us in so many ways. They

have complex societies, forge long-lasting relationships, feel pain, are cognitive and able to resolve
problems, and live freely in an environment where
they can cover great distances and communicate
with each other. Should we accord big-brained, sen-
tient mammals the same basic rights of freedom
that we accord humans?

Knowing what I know as a scientist, I can't
avoid comparing what I study in the wild with the
confined world of trained dolphins. When I think
of how a dolphin might perceive its surroundings
in captivity, I think about what Mara saw when she
approached the porthole of her tank to stare back
at me. Did she understand that I was free while
she was not? I think about the school of acrobatic
bottlenose dolphins I followed in the wild just a
few days ago, free to move in a fluid world of virtually unbridled dimension.

Many conservationists believe that a century from
now, great apes will live only in a few carefully protected nature sanctuaries and in captivity. If that is

the case, then we need to think proactively about how to manage and care for such captive populations. The needs of great apes in captivity are not unlike those of children—they require nurturing, psychological and emotional enrichment, and healthy socialization. The population would also need to be carefully managed genetically, as most species already are. Zoos swap gorillas and orangutans back and forth according to guidelines set by committees that are the keepers of studbooks, with an eye toward avoiding inbreeding and promoting the genetic health of the species.

But can we really keep great apes alive as a captive species indefinitely? Many would say no, exactly because of the profound needs of these animals. The average zoo does not have the space or funds to properly maintain more than one great ape species. As zoos have become enlightened about ethical standards of animal welfare, questions have been raised about whether chimpanzees and their kin ever belong in captive enclosures. As we learn more about their humanlike cognition, it becomes clear that what we never do to a severely cognitively disabled child or adult we do routinely to great apes of similar mental stature. Perhaps the

best argument for maintaining great apes in captivity is the default one—those that have been born
in zoos and laboratories can never be successfully
returned to the wild, and therefore compose the
nucleus of the species' captive population for the
future.

Are There Solutions?

In the face of such bleak prospects, we must assume that in the coming centuries, some species of
wild animals will become extinct, at least in their
natural wild state, and many others will live on
only as highly managed species in well-protected
areas. Just as Americans converted the vast Great
Plains to farmland with a century of rapid population expansion and growth, the African continent
will be converted to agriculture and towns, sweep-
ing away most of the remaining rain forests and
their inhabitants. The goal for conservationists is
to preserve viable, self-sustaining populations of
animals in tracts of habitat as large as can be preserved. In some cases, we can try to maintain small
areas of protected habitat with links, or corridors
of habitats between them.

But how is such preservation to be accomplished? Experience has shown that one key factor in protecting a forest or a species is making it clear to local people that they have a stake in the future of the place and its animals. In Uganda, people living around the national parks in which the lives of the last mountain gorillas hang by a thread have largely embraced the need to protect them. This is not because of national pride and admiration for the animals—these are low priorities when your children are sick or hungry. Economic incentives that better peoples' lives are the key. The Ugandans living around Bwindi Impenetrable National Park receive a small percentage of the revenue that comes from gorilla ecotourism in the form of funds used to build or improve health clinics and to hire doctors. The dollar or two a day to be earned selling a few sodas to a passing tourist may be enough added income for a family to think twice before illegally felling trees in the same forest that is now providing the family with income.

This sort of financial balance of ecotourism and wildlife conservation exists in a fragile state in various developing countries. It is as fragile as the government of that country is unstable, generally. The first sign of a coup or civil war—both frequent occurrences in Africa—brings tourism to a halt, and the incentive for local people to protect the animals disappears. Tourism also has its downside for the apes themselves. Since gorillas and chimpanzees catch nearly all human diseases, the same tourists whose dollars help to save the animals occasionally sneeze their germs close enough to a gorilla to kill it. Great apes that are watched closely by people have been afflicted with respiratory ailments, polio, tuberculosis, and the common cold. Although we can rarely show a direct mode of transmission from human to ape, it seems likely that some of these outbreaks have been caused by people.

For the great apes, the solutions are complicated. Curtailing the bush-meat trade means persuading local African governments to ban the sale of ape meat in markets and arresting people found transporting or selling it. But although policies can be announced, enforcement is another issue. Ending the bush-meat trade means changing the cultural values of Africans. This is like training Americans not to expect turkey at Thanksgiving; it is no easy task. But many conservationists are at work on the problem, trying to find substitute meat sources that people will adopt.

We must hope that the countries in which wild apes live care enough about their natural heritage to preserve them for future generations. Ultimately, whether or not to save them is up to Africans, not the West. A major hope is that better education in Africa will lead African scholars to work on these intractable problems. They can much more effectively connect with local people and local governments. Some countries lag far behind, but others have already produced a generation of talented scientists who are hard at work on solutions to the great ape extinction crisis. Making conservation a culture of the host country, rather than a new form of colonialism imposed by the West, can be achieved only with the help of such local expertise.

The situation for cetaceans is far worse than for apes. With rare exceptions, they do not live within the territorial boundaries of any one country. We can't save dolphins and whales simply by protecting an area of the seas. Unlike land-based national parks and sanctuaries, sea-based ones are not pro-tected islands. This means negotiating interna-tional treaties every time a new conservation issue arises. The oceans are expansive entities where so many things happen out of sight—from illegal dumping to illegal fishing—that enforcement of pro-tection policies is extremely difficult.

Some solutions exist. To prevent bycatch, for example, fishing gear can be modified to allow nontarget species a means of escape, or acoustic "pingers" may be attached to nets so that dolphins are not caught in the first place. Also, over 9 mil-lion tourists a year go whale-watching, making this business as valuable as whaling from an economic standpoint. Could one become the substitute for the other? It would require significant incentives and re-education, but perhaps this is another op-portunity for partial solutions worth exploring. Where there is economic gain to be had or rechan-neled, therein lies the possibility for effecting changes in long-standing traditions and beliefs that may inhibit conservation. Money talks, and we need to look for situations in which creative thinking will produce opportunities.

We must continue to seek solutions, even if we

falter in the short term. To protect dolphins, we must first learn how to protect and manage the ecosystems in which they live—those ecosystems that, in many cases, we have already brought to the brink of destruction. Without protecting their home, there is no hope of saving the dolphins. If we eat their food, we need to find a way to do it that doesn't irreparably deplete the available resources. We must learn how to intelligently co-exist with other species on this planet. But that's not all.

If we take the common management approach that seeks to understand and adjust for short-term trends in any given corner of the food web, we probably overlook the history of the thing we are trying to manage. Consider blue whale populations, for example. We need to protect these animals because they are in danger of extinction in many parts of the world, so we enact legislation and controls that attempt to accomplish this. But in failing to consider the historical view of blue whale stocks, we are effectively basing our conservation efforts on a concept of "normal" that does not take the historical decline into consideration and does not give us an accurate idea of what caused the problem in the first place.

Restoring the natural status of the ocean to whatever it may have been in the past is most likely an impossible task, due to the sheer magnitude of the problems that face it. Unfortunately the ecosystem collapses occurring in many parts of the world may not be reversible and represent cases in which we can never go back. Scientists and legislators need to turn their attention to making sure the damage is assessed properly and take the short-term steps necessary to prevent more damage, and we all need to learn from our mistakes.

Strengthening and locally enforcing fishing and whaling regulations, raising public awareness through education on sustainable fisheries and ocean pollution, increasing research efforts, understanding ecological relationships and linkages in the marine environment, reducing fishery quotas, modifying fishing gear, and developing marine protected areas and time-area closures for fishing are just some of the efforts needed to bring about change.

In all conservation efforts, we need to watch our potential solutions carefully. We need to make sure that our solutions do not wind up contributing to our problems. The precautionary principle dictates that "if the potential consequences of an action may be severe or irreversible, in the absence of full scientific certainty the burden of proof falls on those who would advocate taking the action." In other words: above all else, do no harm. If we are to change the current trends and begin to reverse some of the environmental devastation that our species has inflicted, we must adopt a precautionary approach that incorporates this principle.

As scientists who have spent many years studying dolphins and apes in the wild, we believe that our research, and that of others like us, must incorporate a respect and sense of stewardship for the animals we study. Without keeping a weather eye toward the conservation and protection of these species and the ecosystems in which they live, they will not survive to see the next century.

FURTHER READING

1. An Eternal Fascination

Anthony Alpers, *Dolphins: the Myth and the Mammal* (Boston: Houghton Mifflin, 1960).

Eleanore Devine and Martha Clark, eds., *The Dolphin Smile* (New York: Macmillan, 1967).

Dian Fossey, *Gorillas in the Mist* (Boston: Houghton Mifflin, 1983).

Jane Goodall, *The Chimpanzees of Gombe: Patterns of Behavior* (Cambridge, MA: Harvard University Press, 1986).

John Lilly, *Man and Dolphin* (New York: Pyramid Books, 1961).

Charles A. Lockwood, William H. Kimbel, and John M. Lynch, "Morphometrics and Hominoid Phylogeny: Support for a Chimpanzee-Human Clade and Differentiation among Great Ape Subspecies," *Proceedings of the National Academy of Sciences* 101(2004): 4356–4360.

Ashley Montagu, "The History of the Dolphin" in Tony Frohoff and Brenda Peterson, eds., *Between Species: Celebrating the Dolphin-Human Bond* (San Francisco: Sierra Club Books, 2003), pp. 27–40.

Everhard J. Slijper, *Whales* (New York: Basic Books, 1962).

2. Two Histories Afield

Michael A. Bigg, Graeme M. Ellis, John K. B. Ford, and Kenneth C. Balcomb, *Killer Whales: A Study of Their Identification, Genealogy, and Natural History in British Columbia and Washington State* (Nanaimo: Phantom Press, 1987).

David K. Caldwell, "Notes on the Spotted Dolphin, *Stenella plagidon,* and the First Record of the Common Dolphin, *Delphinus delphis,* in the Gulf of Mexico," *Journal of Mammalogy* 36(1955): 467–470.

David K. Caldwell and Melba C. Caldwell, "Epimeletic (Care-giving) Behavior in Cetacea" in Kenneth Norris, ed., *Whales, Dolphins and Porpoises* (Berkeley: University of California Press, 1966), pp. 755–789.

Richard C. Connor, Rachel A. Smolker, and Andrew F. Richards, "Dolphin Alliances and Coalitions" in Alexander H. Harcourt and Frans B. M. de Waal, eds., *Coalitions and Alliances in Humans and Other Animals* (Oxford: Oxford University Press, 1992), pp. 415–443.

Takeshi Furuichi, "Sexual Swelling, Receptivity, and Grouping of Wild Pygmy Chimpanzee Females at Wamba, Zaïre," *Primates* 28(1987): 309–318.

Takayoshi Kano, *The Last Ape* (Stanford: Stanford University Press, 1992).

Cheryl D. Knott, "Changes in Orangutan Caloric Intake, Energy Balance, and Ketones in Response to Fluctuating Fruit Availability," *International Journal of Primatology* 19(1998): 1029–1043.

Janet Mann, Richard C. Connor, Peter L. Tyack, and Hal Whitehead, eds., *Cetacean Society: Field Studies of Dolphins and Whales* (Chicago: University of Chicago Press, 2000).

Arthur F. McBride and Donald O. Hebb, "Behavior of the Captive Bottlenose Dolphin *Tursiops truncatus,*" *Journal of Comparative Psychology* 41(1948): 111–123.

Kenneth S. Norris, John H. Prescott, Paul V. Asa-Dorian, and Paul Perkins, "An Experimental Demonstration of Echolocation Behavior in the Porpoise, *Tursiops truncatus,* Montagu," *Biological Bulletin* 120(1961): 163–176.

Kenneth S. Norris, Bernd Würsig, Randall S. Wells, and Melany Würsig, *The Hawaiian Spinner Dolphin* (Berkeley: University of California Press, 1991).

Carel P. Van Schaik, *Among Orangutans* (Cambridge, MA: Harvard University Press, 2004).

Randall S. Wells, "Structural Aspects of Dolphin Societies" (Ph.D. diss., University of California, Santa Cruz, 1986), p. 234.

3. Swimming with Dolphins, Swinging with Apes

Robin W. Baird, "The Killer Whale: Foraging Specializations and Group Hunting" in Mann, Connor, Tyack, and Whitehead, eds., *Cetacean Society*, pp. 127–153.

Maddalena Bearzi, "California Sea Lions Use Dolphins to Locate Food," *Journal of Mammalogy* 87(2006): 606–617.

Christophe Boesch and Hedwige Boesch, "Hunting Behavior of Wild Chimpanzees in the Taï National Park," *American Journal of Physical Anthropology* 78(1989): 547–573.

Richard C. Connor, Randall S. Wells, Janet Mann, and Andrew Read, "The Bottlenose Dolphin: Social Relationships in a Fission-Fusion Society" in Mann, Connor, Tyack, and Whitehead, eds., *Cetacean Society*, pp. 91–126.

Stephen Leatherwood and Randall R. Reeves, eds., *The Bottlenose Dolphin* (San Diego: Academic Press, 1990).

John C. Mitani, David Watts, and Jerry Lwanga, "Ecological and Social Correlates of Chimpanzee Party Size and Composition" in Christophe Boesch, Gottfried Hohmann, Linda F. Marchant, eds., *Behavioral Diversity in Chimpanzees and Bonobos* (Cambridge: Cambridge University Press, 2002), pp. 248–258.

Jill D. Pruetz, "Use of Caves by Savanna Chimpanzees in Senegal," *Pan Africa News* (2002).

Jill D. Pruetz and Paco Bertolani, "Savanna Chimpanzees, *Pan troglodytes verus,* Hunt with Tools," *Current Biology* 17(2007): 1–6.

John E. Reynolds III and Sentinel A. Rommel, eds., *Biology of Marine Mammals* (Washington, DC: Smithsonian Institution Press, 1999).

John E. Reynolds III, Randall S. Wells, and Samantha D. Eide, eds., *The Bottlenose Dolphin: Biology and Conservation* (Gainesville: University Press of Florida, 2000).

Jordi Sabater Pi, Magdalena Bermejo, Germain Ilera,

and Joaquim J. Vea, "Behavior of Bonobos *(Pan paniscus)* Following Their Capture of Monkeys in Zaïre," *International Journal of Primatology* 14(1993): 797–804.

Craig B. Stanford, "The Hunting Ecology of Wild Chimpanzees: Implications for the Behavioral Ecology of Pliocene Hominids," *American Anthropologist* 98(1996): 96–113.

———, "The Social Behavior of Chimpanzees and Bonobos: Empirical Evidence and Shifting Assumptions," *Current Anthropology* 39(1998): 399–420.

———, *Chimpanzee and Red Colobus: The Ecology of Predator and Prey* (Cambridge, MA: Harvard University Press, 1998).

4. Dolphin and Ape Societies

Robin W. Baird, "The Killer Whale: Foraging Specializations and Group Hunting" in Mann, Connor, Tyack, and Whitehead, eds., *Cetacean Society,* pp. 127–153.

Richard C. Connor and Michael Kruetzen, "Levels and Patterns in Dolphin Alliance Formation" in Frans B. M. de Waal and Peter L. Tyack, eds., *Animal Social Complexity: Intelligence, Culture, and Individualized Societies* (Cambridge, MA: Harvard University Press, 2003), pp. 115–120.

Richard C. Connor, Randall S. Wells, Janet Mann, and Andrew Read, "The Bottlenose Dolphin: Social Relationships in a Fission-Fusion Society" in Mann, Connor, Tyack, and Whitehead, eds., *Cetacean Society,* pp. 91–126.

Frans B. M. de Waal, "Tension Regulation and Nonreproductive Functions of Sex in Captive Bonobos *(Pan paniscus),*" *National Geographic Research Reports* 3(1987): 318–335.

Frans B. M. de Waal and Frans Lanting, *Bonobo: The Forgotten Ape* (Berkeley: University of California Press, 1997).

Michael Kruetzen, William B. Sherwin, Richard C. Connor, Lynne M. Barre, Tom Van de Casteele, Janet Mann, and Robert Brooks, "Contrasting Relatedness Patterns in Bottlenose Dolphins *(Tursiops* sp.) with Different Alliance Strategies," *Proceedings of the Royal Society of London B* 270(2003): 497–502.

William C. McGrew, Linda F. Marchant, and Toshisada Nishida, *Great Ape Societies* (Cambridge: Cambridge University Press, 1996).

Karen Pryor and Kenneth S. Norris, eds., *Dolphin Societies: Discoveries and Puzzles* (Berkeley: University of California Press, 1991).

Craig B. Stanford, *Significant Others* (New York: Basic Books, 2001).

Richard W. Wrangham, Colin A. Chapman, Adam P.

Clark-Arcadi, and Gilbert Isabirye-Basuta, "Social Ecology of Kanyawara Chimpanzees: Implications for Understanding the Costs of Great Ape Groups" in William C. McGrew, Linda F. Marchant, and Toshisada Nishida, eds., *Great Ape Societies* (Cambridge: Cambridge University Press, 1996), pp. 45–57.

5. Cognition

David E. Bain, "Acoustic Behavior of *Orcinus:* Periodicity, Sequences, Correlations with Behavior, and an Automated Technique for Call Classification" in Barbara C. Kirkevold and Joan S. Lockard, eds., *Behavioral Biology of Killer Whales* (New York: Liss, 1986), pp. 335–371.

Benjamin B. Beck, *Animal Tool Behavior: The Use and Manufacture of Tools by Animals* (Washington, DC: Taylor and Francis, 1980).

Richard W. Byrne, *The Thinking Ape* (Oxford: Oxford University Press, 1995).

Richard Byrne and Anne Russon, "Learning by Imitation: A Hierarchical Approach," *Behavioral and Brain Sciences* 21(1998): 667–721.

Richard W. Byrne and Andrew Whiten, eds., *Machiavellian Intelligence* (Oxford: Clarendon Press, 1988).

Wendi Fellner, Gordon B. Bauer and Heidi E. Harley,

"Cognitive Implications of Synchrony in Dolphins," *Aquatic Mammals* 32(2006): 511–516.

John K. B. Ford, "Vocal Traditions among Resident Killer Whales *(Orcinus orca)* in Coastal Waters of British Columbia," *Canadian Journal of Zoology* 69(1991): 1454–1483.

Roger Fouts and Steven T. Mills, *Next of Kin* (New York: William Morrow, 1997).

Marc D. Hauser, Cory T. Miller, Kathy Liu, and Renu Gupta, "Cotton-top Tamarins *(Saguinus oedipus)* Fail to Show Mirror-Guided Self-Exploration," *American Journal of Primatology* 53(2001): 131–137.

Luis M. Herman, ed., *Cetacean Behavior: Mechanisms and Functions* (New York: John Wiley and Sons, 1980).

Luis M. Herman, Douglas G. Richards, and James P. Wolz, "Comprehension of Sentences by Bottlenosed Dolphins," *Cognition* 16(1984): 129–219.

Vincent M. Janik, "Whistle Matching in Wild Bottlenose Dolphins *(Tursiops truncatus),*" *Science* 289(2000): 1355–1357.

Michael Kruetzen, Janet Mann, Michael R. Heithaus, Richard C. Connor, Lars Bejder, and William B. Sherwin, "Cultural Transmission of Tool Use in Bottlenose Dolphins," *Proceedings of the National Academy of Sciences* 105(2005): 8939–8943.

Stan A. Kuczaj II, Robin D. Paulos and Joana A. Ramos, "Imitation in Apes, Children and Dolphins: Implications for the Ontogeny and Phylogeny of Symbolic Representation" in Laura Namy, ed., *Symbol Use and Symbolic Development* (Cambridge, MA: MIT Press, 2005), pp. 221–243.

John Lilly, *Man and Dolphin* (New York: Pyramid Books, 1961).

Diana Reiss and Lori Marino, "Mirror Self-Recognition in the Bottlenose Dolphin: A Case of Cognitive Convergence," *Proceedings of the National Academy of Sciences* 98(2001): 5937–5942.

Diana Reiss, Brenda McCowan, and Lori Marino, "Communicative and Other Cognitive Characteristics of Bottlenose Dolphins," *Trends in Cognitive Sciences* 1(1997): 123–156.

Luke Rendell and Hal Whitehead, "Culture in Whales and Dolphins," *Behavioural and Brain Science* 24(2001): 309–382.

John E. Reynolds III and Sentinel A. Rommel, eds., *Biology of Marine Mammals* (Washington, DC: Smithsonian Institution Press, 1999).

Sue Savage-Rumbaugh and Roger Lewin, *Kanzi: The Ape at the Brink of the Human Mind* (New York: John Wiley & Sons, 1994).

Rachel Smolker, Andrew Richards, Richard Connor, Janet Mann, and Per Berggren, "Sponge Carrying by Dolphins (*Delphindea, Tursiops* sp.): A Foraging Specialization Involving Tool Use?" *Ethology* 103(1997): 454–465.

C. K. Taylor and Graham S. Saayman, "Imitative Behaviour of Indian Ocean Bottlenose Dolphins *(Tursiops aduncus)* in Captivity," *Behaviour* 44(1973): 286–297.

Michael Tomasello, "Do Apes Ape?" in Cecilia Heyes and Bennett Galef Jr., eds, *Social Learning in Animals: The Roots of Culture* (New York: Academic Press, 1996), pp. 319–346.

Michael Tomasello and Josep Call, *Primate Cognition* (Oxford: Oxford University Press, 1997).

Peter L. Tyack, "Dolphins Whistle a Signature Tune," *Science* 289(2000): 1310–1311.

Elisabetta Visalberghi and Luca Limongelli, "Lack of Comprehension of Cause-Effect Relationships in Tool-Using Capuchin Monkeys *(Cebus apella),*" *Journal of Comparative Psychology* 108 (1994): 15–22.

Andrew Whiten and Richard W. Byrne, *Machiavellian Intelligence II: Extensions and Evaluations* (Cambridge: Cambridge University Press, 1997).

6.   Master Politicians

Richard W. Byrne and Andrew Whiten, "Towards the Next Generation in Data Quality: A New Survey of Primate Tactical Deception," *Behavioral and Brain Sciences* 11(1988): 267–273.

——, "Cognitive Evolution in Primates: The 1990 Database," *Primate Report* 27(1992): 1–101.

Richard C. Connor, Randall S. Wells, Janet Mann, and

Andrew Read, "The Bottlenose Dolphin: Social Relationships in a Fission-Fusion Society" in Mann, Connor, Tyack, and Whitehead, eds., *Cetacean Society,* pp. 91–126.

Jacques-Yves Cousteau and Philippe Diole, *Dolphins* (New York: Arrowood Press, 1987), pp. 304.

Frans B. M. de Waal, "The Chimpanzee's Service Economy: Food for Grooming," *Evolution and Human Behavior* 18(1997): 375–386.

Marc D. Hauser, "Minding the Behaviour of Deception" in Andrew Whiten and Richard W. Byrne, eds., *Machiavellian Intelligence II: Extensions and Evaluations* (Cambridge: Cambridge University Press, 1997), pp. 112–143.

Stan Kuczaj, Karissa Tranel, Marie Trone, and Heather Hill, "Are Animals Capable of Deception or Empathy? Implications for Animal Consciousness and Animal Welfare," *Animal Welfare* 10(2001): 161–173.

David Lusseau, "The Emergent Properties of a Dolphin Social Network," *Proceedings of the Royal Society of London B* (270, 2003): 186–188.

David Lynch and Paul L. Kordis, *Strategy of the Dolphin: Scoring a Win in a Chaotic World* (New York: Fawcett Columbine, 1988), pp. 287.

Janet C. Mann, Richard C. Connor, Lynne M. Barre, and Mike Heithaus, "Female Reproductive Success in Bottlenose Dolphins (*Tursiops* sp.): Life History,

Habitat, Provisioning, and Group-Size Effects," *Behavioral Ecology* 11(2000): 210–219.

Pliny, the Elder, *Natural History,* vol. III, book 9, Loeb Classical Library (Cambridge, MA: Harvard University Press), pp. 177–187.

Craig B. Stanford, Janette Wallis, Eslom Mpongo, and Jane Goodall, "Hunting Decisions in Wild Chimpanzees," *Behaviour* 131(1994): 1–20.

Andrew Whiten and Richard W. Byrne, *Machiavellian Intelligence II: Extensions and Evaluations* (Cambridge: Cambridge University Press, 1997).

Richard W. Wrangham, "Evolution of Coalitionary Killing," *Yearbook of Physical Anthropology* 42(1999): 1–30.

Richard W. Wrangham and Dale Peterson, *Demonic Males* (Boston: Houghton-Mifflin, 1996).

7.   Culture Vultures

Ian Anderson, "Dolphins Pay High Price for Handouts," *New Scientist* 1947(1994): 5.

David E. Bain, "Acoustic Behavior of *Orcinus:* Periodicity, Sequences, Correlations with Behavior, and an Automated Technique for Call Classification," in Barbara C. Kirkevold and Joan S. Lockard, eds., *Behavioral Biology of Killer Whales* (New York: Liss, 1986), pp. 335–371.

John K. B. Ford, "Vocal Traditions among Resident Killer Whales *(Orcinus orca)* in Coastal Waters of British Columbia," *Canadian Journal of Zoology* 69(1991): 1454–1483.

Christophe Guinet and Jerome Bouvier, "Development of Intentional Stranding Hunting Techniques in Killer Whales *(Orcinus orca)* Calves at Crozet Archipelago," *Canadian Journal of Zoology* 73(1995): 27–33.

Vincent M. Janik, "Use of Mimicry on a Dolphin Community in the Wild," *Science* 289(2000): 1355–1357.

Vincent M. Janik, Laela S. Sayigh, and Randall S. Wells, "Signature Whistle Shape Conveys Identity Information to Bottlenose Dolphins," *Biological Sciences* 103(21, 2006): 8293–8297.

William C. McGrew, *Chimpanzee Material Culture* (Cambridge: Cambridge University Press, 1992).

Scott Norris, "Creatures of Culture? Making the Case for Cultural Systems in Whales and Dolphins," *Bioscience* 52(2002) 1: 9–14.

Greg M. O'Corry-Crowe, Robert S. Suydam, Andrew Rosenberg, Kathy J. Frost and Andrew E. Dizon, "Phylogeography, Population Structure and Dispersal Patterns of the Beluga Whale *Delphinapterus leaucas* in the Western Nearctic Revealed by Mitochondrial DNA," *Molecular Ecology* 6(1997): 955–970.

Karen W. Pryor, John Lindbergh, Scott Lindbergh, and Raquel Milano, "A Dolphin-Human Fishing Cooperative in Brazil," *Marine Mammal Science* 6(1990): 77–82.

Luke Rendell and Hal Whitehead, "Culture in Whales and Dolphins," *Behavioural and Brain Science* 24(2001): 309–382.

Rachel Smolker, Andrew Richards, Richard Connor, Janet Mann, and Per Berggren, "Sponge Carrying by Dolphins (*Delphindae, Tursiops* sp.): A Foraging Specialization Involving Tool Use? *Ethology* 103 (1997): 454–465.

Andrew Whiten, Jane Goodall, William C. McGrew, Toshisada Nishida, Vernon Reynolds, Y. Sugiyama, Caroline E. G. Tutin, Richard W. Wrangham, and Christoph Boesch, "Cultures in Chimpanzees," *Nature* 399(1999): 682–685.

8. Toward the Roots of Human Intelligence

Giovanni Bearzi, Stefano Agazzi, Silvia Bonizzoni, Marina Costa, and Arianna Azzellino, "Dolphins in a Bottle: Abundance, Residency Patterns and Conservation of Bottlenose Dolphins *Tursiops truncatus* in the Semi-closed Eutrophic Amvrakikos Gulf, Greece," *Aquatic Conservation: Marine and Freshwater Ecosystems* 17(2007).

Robin I. M. Dunbar, "Neocortex Size as a Constraint on Group Size in Primates," *Journal of Human Evolution* 20(1992): 469–493.

Robert A. Foley and Phyllis C. Lee, "Finite Social Space, Evolutionary Pathways and Reconstructing Hominid Behavior," *Science* 243(1989): 901–906.

Patrick R. Hof, Rebecca Chanis, and Lori Marino, "Cortical Complexity in Cetacean Brains," *The Anatomical Record Part A* 287A(2005): 1142–1152.

Donald C. Johanson, Owen C. Lovejoy, William H. Kimbel, Tim D. White, Steven C. Ward, Michael E. Bush, Bruce M. Latimer, and Yves Coppens, "Morphology of the Pliocene Partial Hominid Skeleton (AL 288-1) from the Hadar Formation, Ethiopia," *American Journal of Physical Anthropology* 57(1982): 403–452.

Margaret Klinowska, "Brains, Behaviour and Intelligence in Cetaceans (Whales, Dolphins and Porpoises)" in Orn D. Jonsson, ed., *Whales and Ethics* (Reykjavik: University of Iceland Press, 1992), pp. 23–37.

Louis Lefebvre, Lori Marino, Daniel Sol, Sebastien Lemieux-Lefebvre, and Saima Arshad, "Large Brains and Lengthened Life History Periods in Odontocetes," *Brain, Behavior and Evolution* 68(2006): 218–228.

Owen C. Lovejoy, "The Evolution of Human Walking," *Scientific American* 259(1988): 118–125.

Lori Marino, "What Can Dolphins Tell Us about Primate Evolution?" *Evolutionary Anthropology* 5(1996): 73–110.

Eduardo Mercado III, Robert K. Uyeyama, Adam A. Pack, and Louis M. Herman, "Memory for Action Events in the Bottlenosed Dolphin," *Animal Cognition* 2(1999): 17–25.

Diana Reiss, Brenda McCowan, and Lori Marino, "Communicative and Other Cognitive Characteristics of Bottlenose Dolphins," *Trends in Cognitive Sciences* 1(4, 1997): 123–156.

John E. Reynolds III and Sentinel A. Rommel, eds., *Biology of Marine Mammals* (Washington, DC: Smithsonian Institution Press, 1999).

Craig B. Stanford, *Upright* (Boston: Houghton-Mifflin, 2003).

"Hans" J. G. M. Thewissen, Martin J. Cohn, Lauren S. Stevens, Sunil Bajpai, John Heyning, and Walter E. Horton Jr., "Developmental Basis for Hind-Limb Loss in Dolphins and Origin of the Cetacean Bodyplan," *Proceedings of the National Academy of Sciences USA* 103(22, 2006): 8414–8418.

Conclusion: Beautiful Minds Are a Terrible Thing to Waste

Giovanni Bearzi, Elena Politi, Stefano Agazzi, Sebastiano Bruno, Marina Costa, and Silvia Bonizzoni, "Occurrence and Present Status of Coastal Dolphins (*Delphinus delphis* and *Tursiops truncatus*) in the Eastern Ionian Sea," *Aquatic Conservation: Marine Freshwater Ecosystems* 15(2005): 243–257.

Marc Bekoff, *Minding Animals: Awareness, Emotions, and Heart* (Oxford: Oxford University Press, 2002).

Elizabeth L. Bennett, Richard G. Ruggiero, Heather E. Eves, Nathalie D. Bailey, and Andrew W. Tobiason, "Case Study 8.2, The Bushmeat Crisis: Approaches for Conservation" in Martha J. Broom, Gary K. Meffe, and C. Ronald Carroll, eds., *Principles of Conservation Biology,* 3rd ed. (Sunderland, MA: Sinauer, 2006).

Heather E. Eves, "The Sustainability of the Bushmeat Trade in Africa and around the Globe: Conflict, Consensus and Collaboration," *Proceedings of IFAW Forum 2005—Wildlife Conservation: In Pursuit of Ecological Sustainability,* International Fund for Animal Welfare.

Tony Frohoff and Brenda Peterson, eds., *Between Species: Celebrating the Dolphin-Human Bond* (San Francisco: Sierra Club Books, 2003).

Jeremy B. C. Jackson, Michael X. Kirby, Wolfgang H. Berger, Karen A. Bjorndal, Louis W. Botsford, Bruce J. Bourque, Roger H. Bradbury, Richard Cooke, Jon Erlandson, James A. Estes, Terence P. Hughes, Susan Kidwell, Carina B. Lange, Hunter S. Lenihan,

John M. Pandolfi, Charles H. Peterson, Robert S. Steneck, Mia J. Tegner, Robert R. Warner, "Historical Overfishing and the Recent Collapse of Coastal Ecosystems," *Science* 293(2001): 629–638.

Frank T. Kenneth, Brian Petrie, Jae S. Choi, and William C. Leggett, "Trophic Cascades in a Formerly Cod-Dominated Ecosystem," *Science* 308(2005): 1621–1623.

Charles J. Moore, Shelly L. Moore, Molly K. Leecaster and Stephen B. Weisberg, "A Comparison of Plastic and Plankton in the North Pacific Central Gyre," *Marine Pollution Bulletin* 42(12, 2001): 120–124.

Elliott A. Norse and Larry B. Crowder, eds., *Marine Conservation Biology: The Science of Maintaining the Sea's Biodiversity* (Washington, DC: Island Press, 2005).

Daniel Pauly, Villy Christensen, Johanne Dalsgaard, Rainer Froese and Francisco F. Torres Jr., "Fishing Down Marine Food Webs," *Science* 279(1998): 860–863.

Craig B. Stanford, "Gorilla Warfare," *The Sciences* (July/August, 1999): 18–23.

Sergi Tudela, Abdelouahed Kai Kai, Francesc Maynou, Mohamed El Andalossi, and Paolo Guglielmi, "Driftnet Fishing and Biodiversity Conservation: The Case Study of the Large-Scale Moroccan Driftnet Fleet Operating in the Alboran Sea (SW Mediterranean)," *Biological Conservation* 121(2005): 65–78.

John R. Twiss Jr. and Randall R. Reeves, eds., *Conservation and Management of Marine Mammals* (Washington, DC: Smithsonian Institution Press, 1999).

ACKNOWLEDGMENTS

THERE ARE FEW VOCATIONS in life as rewarding as the study of animal behavior, and even fewer with so many hurdles to overcome to achieve a measure of success. Ironically, even though the central focus of one's work is the animals them-selves, it is the ever-present human element that tends to dictate success or failure. At every step, one needs assistance and support from government officials, funding benefactors and grants directors, the local populace, colleagues, friends and allies, and family. We are both deeply grateful for decades of help from all of these.

Maddalena thanks...

In the academic world I wish to thank Dr. William Hamner, Dr. Michael Scott, and Dr. John Heyning, who recently passed away and will be deeply missed. I thank my dear friend Dr. Dan Blumstein for his suggestions and powerful insights on the final stages of the book and for our engaging discussions over numerous homemade dinners.

I wish to thank all the grantors and organizations that have supported my research in various parts of the world, including Ocean Conservation Society, Europe Conservation, Tethys Research Institute, The UCLA Mentor Research Program Fellowship, The Coastal Environmental Quality Initiative Fellowship, The American Cetacean Society, the Natural History Museum of Los Angeles County, and the Santa Monica Bay Restoration Foundation.

My sincere appreciation goes to my research staff and volunteers—many of whom have become dear friends: Paul Ahuja, Joanna Arlukiewicz, Andrea Bachman, Celia Barroso, Jennifer Bass, Andrea Cardinali, Travis Davis, Dr. Jon Feenstra, Alice Hwang, Karyn Jones, Bettina Lynch, Mallory Mattox, Dr. Peter Mendel, Michael Navarro, Lisa Openshaw, Daphne Osell, Silvia Papparello, Shana Rapoport, Monica Varallo, and all the others who helped with the Yucatan Project and the Los Angeles Dolphin Project over the last eighteen years. Thanks also to my friends and colleagues Fabrizio Borsani, Enrique Duhne, and Dr. Daniela Maldini.

Close to my heart, my husband Charlie—to whom this book is dedicated—has been my best friend, my companion in the field and at home, the inspiration and support for my work, and my personal editor. He spent countless days reading, reviewing, and improving this book so that an English reader could make some sense of my Italian native tongue. Without him this book would never have found its place, and to him goes my deepest gratitude and love.

Great recognition goes to my brother Giovanni—a cetologist like myself—for all of the times he carefully reviewed my drafts, to my parents for instilling in me a profound love and enthusiasm for nature since childhood, and to my "aunt" Marcy for her close friendship and support.

A special thanks to my co-author Dr. Craig Stanford, a friend and patient colleague, for his excellent and inspired work on this book.

Craig thanks...

Since the early 1990s and through two separate field studies of great apes, I am grateful to the local people who have worked as field assistants or park wardens, without whom the research would not have been possible. Moreover, many of them have been close friends and confidantes during months of isolation in African forests. I thank Bwindi Chief Warden Keith Masana, Evarist Mbonigaba, Chief Warden Christopher Oreyema, and Gervase Tumwebaze; Gombe Chief Wardens Peter Msuya, Stephan Qolli, and Dattomax Selenyika; Alnazir Haji Mohammed, John Dota, Eslom Mpongo, Hamisi Mkono, Yahaya Almasi, Selemani Yahaya, Gabo Paulo, Bruno Herman, the late Msafiri

Katoto, Issa Salala, the late David Mussa, Karoli Alberto, Hilali Matama, Tofficki Mikidaddi, Madua Juma, Nasibu Sadiki, and Methodi Vyampi. Dr. John Bosco Nkurunungi was a great help during his student days in Bwindi.

I am grateful for the generous funding that has allowed me to conduct my research, especially from the National Geographic Society, the Wenner Gren Foundation for Anthropological Research, the Fulbright Foundation, the L.S.B. Leakey Foundation, the National Science Foundation, Primate Conservation, Inc., and the Jane Goodall Research Center of the University of Southern California. The government of Uganda, specifically the Ugandan Wildlife Authority and the Uganda National Council for Science and Technology, have provided permission to work in Uganda. The offices of Tanzania National Parks and the Serengeti Wildlife Research Institute provided permission for field research in Tanzania.

Also of immense help have been the community of expatriate colleagues who helped in both official and unofficial ways: Drs. Anthony Collins and Janette Wallis at Gombe; Mitchell Keiver and

Drs. Alastair McNeilage, Richard Malenky, Martha Robbins, and Nancy Thompson-Handler in Bwindi.

For their long-term constructive and formative role in patiently reading and editing almost everything I write, I am most grateful to Drs. William C. McGrew, John S. Allen, Christopher Boehm, Nayuta Yamashita, Erin Moore, and many others.

And on the home front, I am, as always, deeply grateful for the support of Erin, Adam, Marika, and Gaelen—my family.

We are both deeply grateful to our editor, Michael Fisher, and to the anonymous reviewers of this book.

INDEX

Aggression: against female chimpanzees, 117; among male chimpanzees, 114, 120, 207, 209, 210, 260; among male dolphins, 22, 120, 187–188; bluff attacks, 113; bullying behavior of chimpanzees, 8; charging displays in chimpanzees, 8, 189–190; female protection against, in whales, 126–127; in dolphins, 123–